Anaerobic Digestion

Edited by J. Rajesh Banu

Published in London, United Kingdom

IntechOpen

Supporting open minds since 2005

Anaerobic Digestion
http://dx.doi.org/10.5772/intechopen.73348
Edited by J. Rajesh Banu
Assistant to the Editor(s): Yukesh Kannah R.

Contributors
Vimala Ebenezer, Tamilarasan K, Demsew Mitiku Teferra, Mario Toledo, Olatunde Samuel Dahunsi, Borja Velazquez-Marti, Orlando Meneses-Quelal, Juan Gaibor-Chávez, Zulay Niño-Ruiz, Anna Sikora, Anna Detman, Damian Mielecki, Aleksandra Chojnacka, Mieczysław Błaszczyk, Anna Schnürer, Maria Westerholm, Adeola Shote, Rajesh Banu, Yukesh Kannah R, Serene Sow Mun Lock, Kok Keong Lau, Azmi Mohd Shariff, Yin Fong Yeong, Norwahyu Jusoh, Na Wu, Yingxiu Zhang, Shunchang Yang, Doan Nguyet, Spyros Svoronos, Pratap Pullammanappallil, Edward Phlips, Cesar M Moreira

First published in London, United Kingdom, 2019 by IntechOpen
IntechOpen is the global imprint of INTECHOPEN LIMITED, registered in England and Wales, registration number: 11086078, The Shard, 25th floor, 32 London Bridge Street
London, SE19SG - United Kingdom
Printed in Croatia

British Library Cataloguing-in-Publication Data
A catalogue record for this book is available from the British Library

Additional hard and PDF copies can be obtained from orders@intechopen.com

Anaerobic Digestion
Edited by J. Rajesh Banu
p. cm.
Print ISBN 978-1-83881-849-4
Online ISBN 978-1-83881-850-0
eBook (PDF) ISBN 978-1-83881-851-7

For EU product safety concerns:
IN TECH d.o.o., Prolaz Marije Krucifikse Kozulić 3, 51000 Rijeka, Croatia, info@intechopen.com or visit our website at intechopen.com.

Meet the editor

Dr. J. Rajesh Banu, Assistant Professor, Department of Civil Engineering, Anna University Regional Campus Tirunelveli, was formerly a postdoctoral fellow and lecturer in the Department of Civil and Environmental Engineering, Sungkyunkwan University, South Korea. He received his PhD from the Centre for Environmental Studies, Anna University, India, in 2006. He also had experience of working in environmental consultancy for one year. He has been a reviewer of many national and international journals and is the author of 160 publications in SCI and SCIE journals with a cumulative impact factor of 550. His publications achieve more than 3000 citations and he has an H-index of 31. He has authored 20 book chapters for international publishers. Dr. Banu obtained two major grants from the Government of India and two consultancies from industries. He has guided 12 PhD students and 60 postgraduate engineering students in the discipline of environmental engineering and science. He was awarded with Young Scientist in an international conference (RACEE-2018) held at SSN college, Chennai, India. Dr. Banu's research centers on biological wastewater treatment, nutrient removal, membrane bioreactor mesophilic and thermophilic high-rate anaerobic treatment of wastewater, microbial fuel cells, energy recovery from waste, and sludge reduction. His prime area of research is the generation of energy from solid waste.

Contents

Preface

The rapid increase in global energy demand has resulted in shortages of fossil fuel supply. Biogas is the most profitable bioenergy generated from first-, second-, and third-generation waste biomass, is the best suitable alternative to fossil fuel, and is cost effective. Biogas consists of approximately 60–65% methane, 25–30% carbon dioxide, and a small amount of toxic gases, etc. Organic waste management via anaerobic digestion is the most commonly preferred technique. During this process the anaerobic microbes play a vital role in conversion of organic waste into biogas. To enhance biogas production, pretreatment prior to anaerobic digestion results in achieving a positive energy yield. Developing a proper system to recover biogas from organic waste will reduce global warming issues. This book intends to provide a comparative overview of a recent update in biogas production, processing, and its application in various sectors. In addition, this book contains several scientific discussions regarding microbes involved in biogas production, processing, and recent technologies for sustainable development. The book provides in-depth information about anaerobic digestion to researchers and graduate students. The editor sincerely thanks all the contributors, whose efforts have brought this book to fruition.

Dr. J. Rajesh Banu
Department of Civil Engineering,
Anna University Regional Campus,
Tirunelveli, India

Section 1

Introduction

Chapter 1

Introductory Chapter: An Overview of Biogas

J. Rajesh Banu and R. Yukesh Kannah

1. Introduction

According to the International Energy Agency Report 2018, the global energy demands (GED) elevated 2.1% from the previous year. However, 70% of GED was met through oil, coal and fossil fuel. Among these, fossil fuel accounts for 81% of total energy demand (TED). The percentage of fossil fuel remains unchanged for the past three decades. Exploitation of fossil fuel extended the emission of carbon dioxide (CO_2) to 32.5 GT (gigatonnes) in the year 2017. Surplus emission of greenhouse gases (GHG) into the atmosphere is the major contributor for global warming and climate change. On considering, the profile of GHG emission researchers comes out with innovative ideas to minimize the emission. Nowadays, researchers and policymakers are working together to recognize alternative energy source to encounter the energy demand and global warming impacts.

Anaerobic digestion (AD) process is the cost-effective and emerging technology to derive biogas from various liquids and solid wastes. AD process is more suitable for valorization of high-strength organic waste under both mesophilic (30–40°C) and thermophilic (50–60°C) conditions. AD process is otherwise termed as biomethanation or biochemical degradation. AD process is a more environmental-friendly, energy-yielding and more efficient bioenergy production method than other waste processing technologies.

AD process dominant by anaerobic microbes, which plays major role in conversion of organic rich waste biomass into two valuable products such as methane and nutrient rich digested/effluent. Anaerobic breakdown of complex organic waste biomass follows four major steps, and these are (i) hydrolysis, (ii) acidogenesis, (iii) acetogenesis and (iv) methanogenesis. **Figure 1** represents the pathway of anaerobic degradation of organic waste.

Among them, hydrolysis is the rate-limiting and first step of AD process. During hydrolysis, complex organics ($C_6H_{10}O_4$) such as protein, carbohydrate and fat are converted into simple digestible amino acids, monosaccharides and fatty acids. Eq. (1) shows that the reaction occurs during hydrolysis phase; enzymes convert the complex organic substrate into simple monomers ($C_6H_{12}O_6$) and hydrogen (H_2) as shown below:

$$C_6H_{10}O_4 + 2H_2O \rightarrow C_6H_{12}O_6 + H_2 \quad (1)$$

Hydrolysis is a very slow process when compared with other steps involved in AD process. Inadequate hydrolysis of organic waste affects the efficiency of AD. In order to increase the rate of hydrolysis, many researchers have adopted various pretreatment methods. Banu and Kavitha [1] have reviewed in detail regarding various pretreatment methods and their effects on anaerobic digestion.

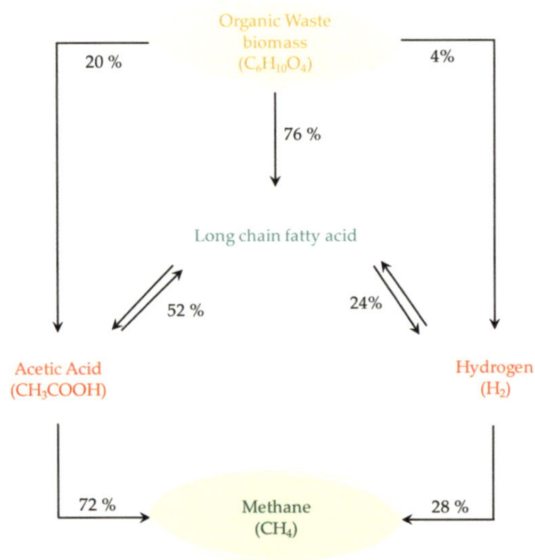

Figure 1.
Pathway of anaerobic degradation of organic waste.

Pretreatment enhances the digestibility of organic substrate, and it is broadly classified into five major groups. They are physical, chemical, biological, mechanical and combinative pretreatments. **Figure 2** shows pretreatment methods and their classification. Physical pretreatment is further classified into two: thermal [2] microwave [3] and freezing and thaw [4] pretreatments. Chemical pretreatment is further classified into two: alkaline [5] and acidic [6] pretreatments. Biological pretreatment is further classified into two: enzyme [7] and fungal [8] pretreatments. Mechanical pretreatment is further classified into two: high-pressure homogenizer [9] and ultrasonic [10] pretreatments. Combinative pretreatment such as thermo-chemo-sonic [11], thermo-chemo-disperser [12], thermo-chemo-ozone [13] and hydrothermal [14] pretreatment, etc. Many researchers have experimentally proven the positive effect of pretreatment on hydrolysis and subsequent biogas production [15].

Acidogenesis is the second step involved in AD process; in this step, acidogenic microbes are responsible for conversion of hydrolyzed organics into ethanol (C_2H_5OH), acetate (CH_3COO^-), hydrogen (H_2), carbon dioxide (CO_2) and other acids (propionic, formic, lactic, butyric, succinic acids). In some cases, amino acids cause formation of ammonia. Eqs. (2)–(4) show that the reaction occurs during acidogenesis phase as shown below:

$$C_6H_{12}O_6 \leftrightarrow 2CH_3CH_2OH + 2CO_2 \tag{2}$$

$$C_6H_{12}O_6 + 2H_2 \leftrightarrow 2CH_3CH_2COOH + 2H_2O \tag{3}$$

$$C_6H_{12}O_6 \rightarrow 3CH_3CH_2OH \tag{4}$$

Acetogenesis is the third step involved in AD process. In this step, acetogenic microbes are responsible for conversion of long-chain fatty acid, volatile fatty acid and alcohols into acetic acid (CH_3COOH), hydrogen (H_2) and carbon dioxide

DOI: http://dx.doi.org/10.5772/intechopen.82198

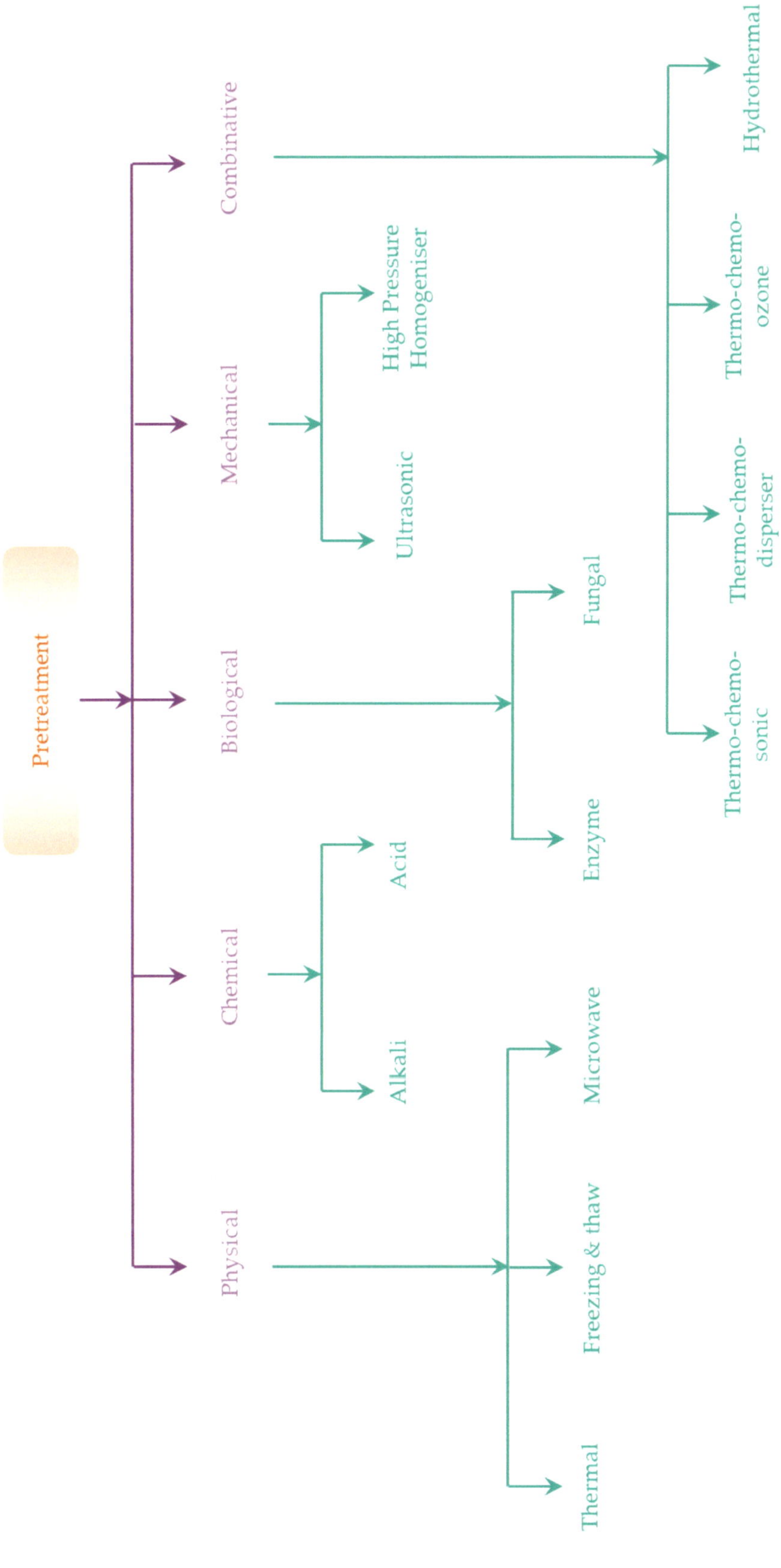

Figure 2.
Pretreatment methods and their classification.

(CO_2). Eqs. (5)–(7) show that the reaction occurs during acetogenesis phase as shown below:

$$CH_3CH_2COO^- + 3H_2O \leftrightarrow CH_3COO^- + H^+ + HCO_3^- + 3H_2 \tag{5}$$

$$C_6H_{12}O_6 + 2H_2O \leftrightarrow 2CH_3COOH + 2CO_2 + 4H_2 \tag{6}$$

$$CH_3CH_2OH + 2H_2O \rightarrow CH_3COO^- + 3H_2 + H^+ \tag{7}$$

During this conversion process, the concentration of biological and chemical oxygen demand in the medium gets reduced. On the other hand, the hydrogen partial pressure is generated due to the presence of hydrogen gas inside the reactor. Methanogenic microbes, present in the digester, consume accumulated hydrogen gas.

Methanogenesis is the final step of anaerobic degradation of organic waste. In this step, methanogenic microbes are responsible for converting the acetic acid and hydrogen into methane (CH_4) gas and carbon dioxide (CO_2). Eqs. (8)–(10) show that the reaction occurs during methanogenesis phase as shown below:

$$CH_3COOH \rightarrow CH_4 + CO_2 \tag{8}$$

$$CO_2 + 4H_2 \rightarrow CH_4 + 2H_2O \tag{9}$$

$$2CH_3CH_2OH + CO_2 \rightarrow CH_4 + 2CH_3COOH \tag{10}$$

Methane-enriched biogas can be a promising source to displace the use of conventional fossil fuel. Biogas acts as a flexible energy source, which can be used for various applications like power, heat, transport and feedstock for chemical production. Biogas is the most significant product of (AD) process, and it comprises 60–70% of methane (CH_4) gas, 25–35% of carbon dioxide (CO_2) and remaining 5–10% of other corrosive gases. Biogas was more suitable to replace the demand of conventional fuel. Biogas has a calorific value of 6.0–6.5 kWh/m^3, which varies according to the percentage of biomethane content in the biogas. In addition to this, AD process indirectly reduces the cost of energy and fuel production. On the other hand, anaerobically digested residues have market value due to its nutrient content. It can be used as bio-fertilizer for agriculture crop production. AD process is termed as a golden process to eliminate the emission of GHG and reduce global warming issues.

According to the World Bioenergy Association 2017 report, the global biomethane production was approximately 35 billion m^3 of methane. Overall, global biogas production was 1.28 EJ in the year 2014. Developed countries like the United States and Europe are the major contributors of biogas production throughout the world. Among them Europe is the world's largest biomethane producer. Around 18 billion m^3 of biomethane was produced in the year 2015; it was half of the global biogas production. The produced biomethane was utilized for generation heat, electricity and transportation (vehicle fuel). Nearly 50% of total biogas was utilized for heat generation, and around 697 biomethane filling stations were employed in Europe [16]. Developing countries in Asia (India,

China, Bangladesh, Pakistan, Sri Lanka and Nepal) and Africa (Burkina Faso, Ethiopia, Tanzania, Kenya and Uganda) are very successful in the operation of domestic scale digester. In Asia, approximately 47.876 million of domestic scale digesters were effectively operated to meet their daily needs. In that, China holds first place and accounts for 43 million domestic scale digester, India 4.75 million, Nepal 330,000, Bangladesh 36,000, Sri Lanka 6000 and Pakistan 4000. Similarly, Africa holds 60,000 domestic scale digesters, in that Kenya leads first place and accounts for 16,419, Ethiopia 13,584, Tanzania 13,037, Uganda 6504 and Burkina Faso 7518. Produced biogas was utilized for cooking and lighting purposes.

Author details

J. Rajesh Banu* and R. Yukesh Kannah
Department of Civil Engineering, Anna University Regional Campus, Tirunelveli, India

*Address all correspondence to: rajeshces@gmail.com

References

[1] Banu JR, Kavitha S. In: Singh L, Kalia VC, editors. Various Sludge Pretreatments: Their Impact on Biogas Generation BT—Waste Biomass Management—A Holistic Approach. Cham: Springer International Publishing; 2017. pp. 39-71

[2] Raj SE, Banu JR, Kaliappan S, Yeom I-T, Adish Kumar S. Effects of side-stream, low temperature phosphorus recovery on the performance of anaerobic/anoxic/oxic systems integrated with sludge pretreatment. Bioresource Technology. 2013;**140**:376-384

[3] Kavitha S, Rajesh Banu J, Kumar G, Kaliappan S, Yeom IT. Profitable ultrasonic assisted microwave disintegration of sludge biomass: Modelling of biomethanation and energy parameter analysis. Bioresource Technology. 2018;**254**:203-213

[4] Zhang X, Chen M, Huang Y. Isothermal drying kinetics of municipal sewage sludge coupled with additives and freeze–thaw pretreatment. Journal of Thermal Analysis and Calorimetry. 2017;**128**:1195-1205

[5] Banu JR, Do Khac U, Kumar SA, Ick-Tae Y, Kaliappan S. A novel method of sludge pretreatment using the combination of alkalis. Journal of Environmental Biology. 2012;**33**:249

[6] Devlin DC, Esteves SRR, Dinsdale RM, Guwy AJ. The effect of acid pretreatment on the anaerobic digestion and dewatering of waste activated sludge. Bioresource Technology. 2011;**102**:4076-4082

[7] Kavitha S, Preethi J, Rajesh Banu J, Yeom IT. Low temperature thermochemical mediated energy and economically efficient biological disintegration of sludge: Simulation and prediction studies for anaerobic biodegradation. Chemical Engineering Journal. 2017;**317**:481-492

[8] Cheng X-Y, Liu C-Z. Fungal pretreatment enhances hydrogen production via thermophilic fermentation of cornstalk. Applied Energy. 2012;**91**:1-6

[9] Fang W, Zhang P, Shang R, Ye J, Wu Y, Zhang H, et al. Effect of high pressure homogenization on anaerobic digestion of the sludge pretreated by combined alkaline and high pressure homogenization. Desalination and Water Treatment. 2017;**62**:168-174

[10] Divyalakshmi P, Murugan D, Sivarajan M, Sivasamy A, Saravanan P, Rai CL. Effect of ultrasonic pretreatment on secondary sludge and anaerobic biomass to enhance biogas production. Journal of Material Cycles and Waste Management. 2018;**20**:481-488

[11] Kavitha S, Yukesh Kannah R, Yeom IT, Do K-U, Banu JR. Combined thermo-chemo-sonic disintegration of waste activated sludge for biogas production. Bioresource Technology. 2015;**197**:383-392

[12] Kavitha S, Jayashree C, Kumar SA, Kaliappan S, Banu JR. Enhancing the functional and economical efficiency of a novel combined thermo chemical disperser disintegration of waste activated sludge for biogas production. Bioresource Technology. 2014;**173**:32-41

[13] Kannah RY, Kavitha S, Rajesh Banu J, Yeom IT, Johnson M. Synergetic effect of combined pretreatment for energy efficient biogas generation. Bioresource Technology. 2017;**232**:235-246

[14] Li C, Wang X, Zhang G, Li J, Li Z, Yu G, et al. A process combining

hydrothermal pretreatment, anaerobic digestion and pyrolysis for sewage sludge dewatering and co-production of biogas and biochar: Pilot-scale verification. Bioresource Technology. 2018;**254**:187-193

[15] Kavitha S, Rajesh Banu J, Subitha G, Ushani U, Yeom IT. Impact of thermo-chemo-sonic pretreatment in solubilizing waste activated sludge for biogas production: Energetic analysis and economic assessment. Bioresource Technology. 2016;**219**:479-486

[16] Scarlat N, Dallemand J-F, Fahl F. Biogas: Developments and perspectives in Europe. Renewable Energy. 2018;**129**:457-472

Section 2

Anaerobic Digestion

Chapter 2

Microbial Responses to Different Operating Practices for Biogas Production Systems

Maria Westerholm and Anna Schnürer

Abstract

Biogas production requires a number of different microbial groups that work in a synchronized and closely interacting manner. For bioreactors constructed to maximize waste treatment and energy production, it is crucial to manage this process in a way that secures the growth and activity of these microorganisms, as otherwise there is a great risk of process failure. However, the microbiome has a remarkable ability to adapt to various conditions related to substrate composition and operating conditions, thus showing high functional redundancy and robustness. In order to optimize and steer the process, it is important to have an understanding of the anaerobic microbiome, how it responds to various conditions, and its upper limits. This chapter reviews current knowledge regarding microbial responses to different operational management strategies. Microbial responses under various conditions and how the process can be operated to maintain the activity of key species are addressed. Parameters discussed include for example substrate composition, pre-treatment, ammonia level, temperature and organic load.

Keywords: anaerobic degradation, microbiology, taxonomy, start-up, temperature, substrate composition, feeding, additives, bioaugmentation

1. Introduction

As the world's population continues to grow, it is necessary to find ways to develop resourceful waste treatment methods while concurrently reducing the dependency on fossil fuels. In this regard, biogas produced through anaerobic degradation (AD) is highly interesting, as it can replace fossil fuels in power and heat production, be used as feedstock for production of biochemicals, or be converted to vehicle fuel [1]. The biogas technology also enables resource sustainability when the digestion residue (digestate) is used as organic fertilizer to replace fossil energy-requiring mineral fertilizers [2].

Anaerobic digestion of organic material to biogas is a complex microbiological process requiring the combined activity of several groups of microorganisms with different metabolic capacities and growth requirements. To obtain a stable and efficient biogas process, it is important to meet the growth requirements of all microorganisms involved. The substrate is one critical parameter in this regard, contributing growth factors and macro- and micronutrients. Some organic materials can be used as the sole substrate, while others have to be co-digested with substrates that are complementary in composition in order to provide favorable conditions for microbial growth [3]. However, addition of additives such as iron, trace metals, or buffering chemicals may be essential in certain processes in order to ensure sufficient microbial activity and to prevent process collapse [4]. In addition to the nutrient composition, operating parameters such as pretreatment method, load of input material, retention time, process temperature, and stirring are of critical importance. All these parameters have to be set at appropriate levels in order to ensure high activity and gas yield with minimized risk of inhibition or washout of critical functions and microorganisms [5–9]. Thus, many different aspects need to be taken into consideration to achieve optimal microbial activity giving a high degree of degradation and gas production. It should be borne in mind that many operating and biological parameters are interlinked, sometimes with counteracting effects.

2. Characteristics of substrates used for biogas production

The composition of substrates can vary considerably between anaerobic digesters, which bring different challenges depending on the feed characteristics combined with the parameters chosen for the specific system. For example, substrates rich in protein and fat have a high energy content and thus a high methane potential, but can sometimes cause process disturbances due to formation of inhibitory compounds or foaming [10–12]. Other materials posing a lower risk of process disturbance, such as lignocellulosic materials, can require an unfeasibly long time for degradation. In order to explain the prerequisites for microbial degradation and the challenges that exist, this section briefly describes the main characteristics of common substrates for biogas production. This provides background for a detailed description of the microbial degradation process and the responses to changes in operating parameters.

Plant-based materials, such as fruit, grains, vegetables, and root crops, are typically rich in different polysaccharides. Polysaccharides are chains of sugars linked in linear chains (cellulose and starch) or branched chains (hemicellulose, pectin, and glycogen). In the plant cell wall, hemicellulose, cellulose, and lignin are associated in the form of lignocellulose [13]. Simple polysaccharides such as starch and glycogen are easily cleaved by microorganisms into glucose units. Hemicellulose and cellulose are also relatively easily degraded but, when combined with lignin (i.e., lignocellulose) as in plants, the structure becomes relatively persistent to microbial degradation [14, 15]. Lignocellulosic materials such as straw (wheat, rice, corn, barley) and sugarcane bagasse are the most abundant renewable biomass and have high potential to contribute to expansion of worldwide biogas production [13, 16].

Protein-rich materials for biogas production include waste from animal rearing (slaughterhouse, dairy, animal manure, aquaculture sludge), ethanol fermentation (distiller's waste), food industry, and households [10, 17–21]. Proteins consist of long chains of amino acids joined by peptide (or amide) bonds and there are 20 different amino acids of various lengths. A feature of all amino acids is that they have at least one amine group ($-NH_2$). The efficiency of protein degradation depends on the structure of these compounds and their solubility [22].

Slaughterhouse waste, food waste, and grease-separation sludge are materials with a high fat content [23–25]. Fat molecules are of different lengths (saturated or unsaturated) and are hydrolyzed to long-chain fatty acids (LCFA, >12 carbon atoms) and glycerol [26]. Lipids are normally rapidly degraded in AD, whereas the conversion of LCFA can represent a rate-limiting step [27, 28].

3. The microbial degradation steps leading to methane

The microbial process comprises the main degradation steps hydrolysis, acidogenesis, acetogenesis, and methanogenesis (**Figure 1**) and this process has to be efficient and balanced in order to obtain successful anaerobic digestion. The initial step is performed by hydrolytic bacteria, and possibly also fungi, that convert polymers (polysaccharides, lipids, proteins, etc.) into soluble monomers (LCFA, glycerol, amino acids, sugars, etc.) [29, 30]. The hydrolytic reaction is mediated by extracellular enzymes secreted by bacteria to the bulk solution and/or attached to their cell wall. Cellulose is hydrolyzed to cellobiose and glucose, while hemicelluloses are degraded to monomeric sugars and acetic acid by bacteria that often have several different enzymes combined into so-called cellulosomes situated on their cell wall [16, 31]. These cellulosomes contain proteins that have the ability to bind to cellulose, which makes the degradation more efficient because the enzymes can work directly "on-site." Fungal cellulases use a different mechanism and not only bind to the surface of the cellulose, but also to penetrate inside the complex biomass materials (e.g., plant cell walls) [32].

Through the action of extracellular enzymes (proteases), proteins are hydrolyzed into amino acids, which are subsequently degraded in the Stickland reaction or through uncoupled oxidation. In the Stickland reaction, one amino acid acts as an electron donor and the other as an electron acceptor, and the oxidation process produces a volatile carboxylic acid that is one carbon atom shorter than the original amino acid. For example, alanine with its three-carbon chain is converted to acetate [33]. Amino acids can also be fermented through uncoupled oxidation where electrons are instead released as hydrogen. This process can only occur in cooperation with a hydrogen-utilizing partner, such as methanogens, that keeps the hydrogen partial pressure low [34]. Irrespective of the degradation pathway, the amino group in the amino acid is released as ammonia and the sulfur in cysteine and methionine results in sulfide. Lipases are excreted by hydrolytic bacteria and catalyze the hydrolysis of lipids at the water-lipid interface [35], forming saturated or unsaturated LCFA and glycerol [36]. LCFAs thereafter absorb to and are transported through microbial

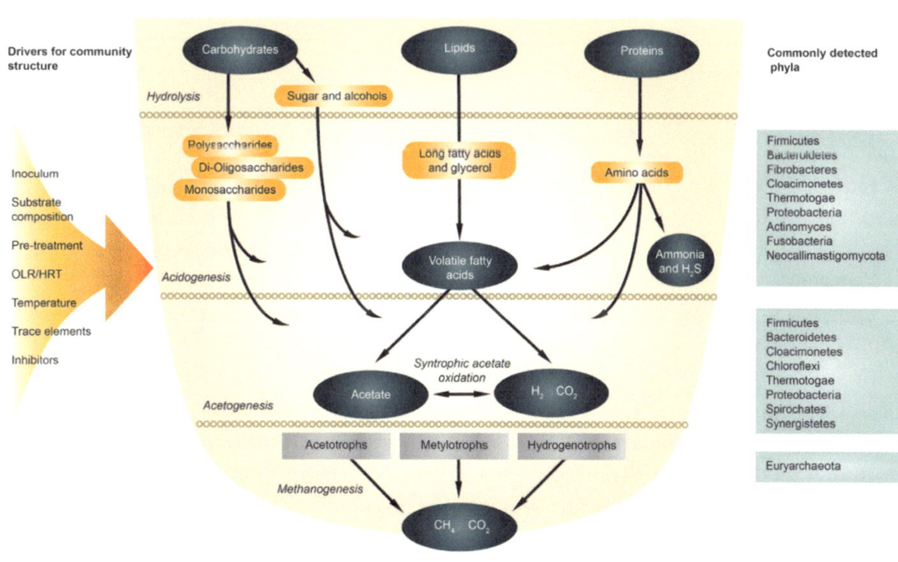

Figure 1.
Anaerobic degradation of carbohydrates, lipids, and proteins and the phyla commonly reported to be involved in the different steps. Biogas digester parameters identified as main drivers for community structure is depicted. The figure is adapted from Kougias et al. [39].

cell membranes of acetogenic bacteria, where the LCFAs are converted to acetate via beta-oxidation to acetate, carbon dioxide (CO_2), and hydrogen (H_2) [37, 38].

The soluble monomers produced in the hydrolytic and acidogenic steps are further degraded to intermediate products. These mainly comprise volatile fatty acids (e.g., acetate, propionate, butyrate, lactate, valerate, and caproate), alcohols, formate, H_2, and CO_2 [40]. During acetogenesis, the products formed in hydrolysis/acidogenesis are further converted by a group of bacteria called acetogens, generating acetate, H_2, and CO_2 as main products. During this process, various electron acceptors can be used, including CO_2, nitrate, sulfate, and protons, with the latter being most important in the biogas process [41]. Acetogens can also directly use products from hydrolysis, such as sugars and amino acids [42], or oxidize pyruvate, which is a common intermediate in anaerobic degradation reactions, to acetate [43]. For thermodynamic reasons, many reactions performed by acetogens, such as oxidation of organic acids and LCFA, can only proceed if the partial pressure of H_2 (p_{H2}) is kept low [44]. For some acids, such as propionate, the removal of acetate can also be of crucial importance [45]. The removal of the acidogenic products acetate and H_2/formate and some methylated compounds mainly proceeds through consumption by methanogens. The energetic situation for the methanogens is comparatively more favorable than acetogenesis, and thus combining these reactions allows both organisms to obtain energy for growth. This type of symbiosis, in which neither organism can operate without the other but together they exhibit metabolic activities that they could not accomplish on their own, is called syntrophy [43, 44].

In the last step, methanogenic archaea use acetate, CO_2, or methylated compounds to produce methane (CH_4) (**Figure 1**). In acetate-utilizing (aceticlastic) methanogenesis, acetate is split into a methyl group and CO_2, and the methyl group is later reduced to methane using an electron provided by the carboxyl group. CO_2 is reduced to methane by hydrogenotrophic methanogens, using H_2 or formate as primary electron donors. In methanogenesis from methylated compounds such as methanol, methylamines, and methylsulfides, the methyl group is reduced to methane. Most methylotrophic methanogens then obtain the electrons they require for reduction from oxidation of additional methyl groups to CO_2 [46, 47].

4. Microorganisms engaged in the different degradation steps

Organisms that are active during the hydrolysis of polysaccharides in biogas processes include various bacteria and anaerobic fungi [14, 29]. Cellulose and starch-degrading bacteria are found within the genera *Acetivibrio*, *Butyrivibrio*, *Caldanaerobacter*, *Caldicellulosiruptor*, *Clostridium*, *Eubacterium*, *Halocella*, *Ruminoclostridium* and *Ruminococcus* (phylum Firmicutes), *Bacteroides* and *Paludibacter* (phylum Bacteroidetes), *Fibrobacter* (phylum Fibrobacteres), Spirochaetes (phylum Spirochaeta), and *Fervidobacterium* and *Thermotoga* (phylum Thermotogae) [14, 48–57]. Identification of the genes necessary for degradation of cellulose has also led to the suggestion that members of the phylum Proteobacteria [56], candidate phylum Cloacimonetes [58] and Actinomyces [59] have this ability. Among the anaerobic fungi, representatives of the phylum Neocallimastigomycota, commonly also found in ruminants, have been suggested as promising candidates to improve biogas production from lignocellulosic material [60, 61]. Protein and amino acid degradation in anaerobic digesters has been shown to be performed by various genera within the phylum Firmicutes, such as *Anaeromusa*, *Anaerosphaera*, *Aminobacterium*, *Aminomonas*, *Gelria*, *Peptoniphilus*,

Thermanaerovibrio [62–67], *Clostridium* [68], *Proteiniborus* [69], and *Sporanaerobacter* [70]. However, members of the phyla Bacteroidetes (e.g., genera *Fermentimonas* and *Proteiniphilum*), Fusobacteria, and Cloacimonetes have also been suggested to have an active amino acid-based metabolism in anaerobic digesters [71, 72]. Less is known about bacteria involved in hydrolysis of fat. Lipolytic bacteria in anaerobic digesters has so far been proposed to belong to families Caldilineaceae (phylum Firmicutes), Bacteroidaceae (phylum Bacteroidetes) and to genera *Trichococcus* (phylum Firmicutes), *Devosia*, and *Psycrobacter* (phylum Proteobacteria) [73, 74].

Acetogenesis and syntrophic acid degradation are often performed by bacteria belonging to the genera *Clostridium* and *Acetobacterium* (phylum Firmicutes), but have also been assigned to the phylum Proteobacteria [14, 43, 75]. Bacteria identified so far that are capable of β-oxidizing LCFA in syntrophy with methanogens all belong to the families Syntrophomonadaceae and Syntrophaceae [23, 76]. Syntrophs that degrade short-chain fatty acids, such as butyrate, propionate, and acetate, in association with methanogens are phylogenetically distributed. Syntrophic propionate and butyrate degradation is performed by genera such as *Syntrophomonas*, *Syntrophospora*, *Syntrophothermus*, *Thermosyntropha*, and *Pelotomaculum* (phylum Firmicutes), or the genera *Syntrophus*, *Smithella*, and *Syntrophobacter* (phylum Proteobacteria) [77]. In addition, the phyla Cloacimonetes, Synergistetes, and Chloriflexi have been suggested to contain bacteria capable of performing syntrophic metabolism in association with hydrogenotrophic methanogens [78–80]. Bacteria capable of syntrophic acetate oxidation identified to date belong to the genera *Clostridium*, *Thermoacetogenium*, *Syntrophaceticus*, and *Tepidanaerobacter* (phylum Firmicutes) [81]. Novel syntrophic acetate-oxidizing bacteria (SAOB) candidates have been suggested within the order Clostridiales and/or Thermoanaerobacterales [82–86], *Synergistes* group 4 [87], the genus *Coprothermobacter* [88] and the phyla Spirochaetes [89], Thermotogae [83], Chloroflexi, and Bacteroidetes [90].

In terms of relative abundance, the methanogenic community generally represents a minor part (2–5%) of the total community, but methanogens have been observed to have high activity relative to their abundance [83, 91, 92]. Methanogens commonly detected in biogas digesters belong to the orders Methanobacteriales, Methanomicrobiales, and Methanosarcinales (phylum Euryarchaeota). However, the orders Methanococcales and Methanomassiliicoccales (phylum Euryarchaeota) have also been found in AD systems [30, 93]. Hydrogenotrophs are found within all methanogenic orders except for the Methanomassiliicoccales [93]. Acetate is only used by members of the families Methanosarcinaceae and Methanosaetaceae (order Methanosarcinales). Members of the Methanosarcinaceae are comparatively more versatile, having the ability to grow on several different substrates, such as acetate, hydrogen, and methanol, while members of the Methanosaetaceae use only acetate [94]. Methane formation from methylated compounds is performed by members of the Methanomassiliicoccales, Methanobacteriales, and Methanosarcinales [93]. A candidate methanogenic class, WSA2, has also been proposed and suggested to be restricted to methanogenesis through methylated thiol reduction [95].

With ongoing advances in molecular techniques and cultivation studies, the list of anaerobic microorganisms responsible for different degradation pathways is continually being updated. The complexity of the cooperation involved in degradation is further illustrated by the fact that members within one and the same genus are often able degrade chemically different compounds. In future, the introduction of omics approaches, combined with isolates of novel species, will most likely increase insights into the taxa involved [30, 96–99].

5. The impact of different operating conditions on AD microbial communities

To optimize the anaerobic digestion process and steer it in a desired direction, it is important to have knowledge and understanding of the metabolic capacities of key microorganisms. Knowledge of the level of functional redundancy within the community (how easily the microbial community adapts to operating changes) and microbial requirements for activity can also help identify operating management practices for improved process performance. In this section, the impact on the microbial community of different operating strategies is described.

5.1 Start-up strategies

The inoculum used for starting up a biogas process has been shown to be of importance for the degradation rate, specific methane yield, and stress tolerance, possibly depending on differences in the composition of the microbial community [52, 100–104]. In addition, chemical parameters, such as presence of trace elements needed for microbial activity, have been suggested to be important [105]. Inocula, most commonly applied in practice, can be categorized as originating from one of the following three sources: wastewater treatment plants, agricultural biogas plants, and plants treating various biowastes, such as municipal and industrial food waste [101]. Microbial analyses of biogas plants belonging to these different groups have clearly shown separation based on microbial community structure [102, 103, 106, 107]. This separation is believed to be caused by the substrate characteristics and operating conditions, with temperature and ammonia being strong regulating parameters [106]. It has been suggested that wastewater sludge is most optimal as the inoculum for biomethane potential (BMP) tests, due to its diverse and highly active community [101]. However, Koch et al. [108] found that inoculum originating from a plant degrading similar substrate to that evaluated in the BMP test gave the best results, suggesting that a substrate-adjusted microbial community is more suitable. Choosing inoculum from a well-functioning biogas process degrading similar substrate and operating under the parameters planned for the new process has also been shown to reduce the period for start-up and avoid initial instability during continuous operation [52, 100]. It has been suggested that methanogenic activity and abundance are appropriate parameters for assessing the suitability of an inoculum and for achieving high rates and yields in BMP tests, as well as for operation of a continuous biogas process [100, 109]. Another factor that can be favorable for the process is to use an inoculum with high microbial diversity, which is considered to correlate with high functional redundancy. One hypothesis to explain this is that having a large number of species provides potential for failing species to be easily replaced by other species performing similar functions, with little impact on the overall process [110].

Evaluations of different inocula during semi-continuous operation using the same substrate have been made for mesophilic processes operating with maize silage [103], a mix of manure and grass [52], cellulose [102], and a mix of waste-activated sludge and glycerol [100]. These studies have produced some contradictory results with regard to the composition of the microbial community over time. Han et al. [102] found that the inoculum source was determining for methane yield, pH, and volatile fatty acid (VFA) production using cellulose as a substrate, both during start-up and after reaching stable operation. Different steady state community patterns were also obtained in the different reactors started with different inocula. Moreover, reactors characterized by high VFA levels and low pH had comparatively low levels of Methanosarcinales, highlighting the importance of this methanogen for efficient biogas production. In line with this, high levels of Methanosarcinales have also been shown to be important for efficient start-up and revival of a thermophilic process suffering from high acetate levels [111]. In contradiction to the results reported by Han et al. [102], a study employing three different inocula for start-up of parallel processes using a manure-grass mix as substrate found that the overall microbial community and process performance became similar in the parallel processes after three hydraulic retention times (HRT) of operation [52]. However, a clear difference in performance was seen during the initial phase after start-up in that study, with poor performance when using an inoculum from a high-ammonia process. Less efficient start-up using a high-ammonia inoculum was also seen in a study by de Vrieze et al. [100] on AD with sludge and glycerol. High-ammonia levels usually impact microbial richness and cause significant shifts in both the bacterial and methanogenic community [82, 106]. This possibly explains the less efficient start-up performance when using substrate with a comparatively low nitrogen level [100]. A negative correlation between ammonia level and cellulose degradation efficiency was also found in the abovementioned study by Liu et al. [52]. Interestingly, when processes started with different inocula and unified in performance and microbial community were supplemented with an additional substrate in that study, the processes again diverged in both performance and microbiology. These results illustrate that choice of inoculum can influence long-term performance of biogas processes [112]. Moreover, even when the same inoculum and operating parameters are used during start-up, different process performances and microbial communities can evolve [102, 113]. This illustrates that stochastic factors play an important role in the microbial community assembly in biogas reactors. It also highlights the need for further research on the impact of inoculum source and operating conditions on long-term effects and optimized performance.

5.2 Temperature

Temperature strongly affects the microbial community structure and thus also process performance and stability [5, 92, 106, 107, 114–118]. When choosing the operating temperature, other operating parameters such as substrate, feeding strategy, and presence or possible formation of inhibitory compounds should be taken into account. The temperatures normally used for digestion in industrial biogas processes are not only mesophilic (37–40°C) or thermophilic (50–55°C), but also psychrophilic (<25°C) and temperatures between mesophilic and thermophilic (41–45°C) have been shown to be achievable [57, 118–122]. Some studies investigating AD at 41–45°C have even reported higher methane production compared with the more commonly used mesophilic or thermophilic range, with associated microbial shifts [57, 119, 121]. In general, metabolic rates and biochemical processes increase with increasing temperature [115, 123, 124]. However, thermophilic conditions can also make the process more sensitive to disturbances and inhibitory compounds [115, 125] and cause less efficient degradation of some inhibitory compounds [126]. Shifts in microbial community in response to temperature change can take time and involve periods of instability. It is therefore recommended to allow the community to adapt to the temperature change by a slow increase/decrease (±1°C per day) [5, 127–130]. In order to avoid process collapse, temperature changes should be carefully monitored, both when increasing and decreasing the operating temperature. A temporary reduction in feed rate and prolonged retention time can be required in the event of disturbance during step-wise temperature changes [5]. Another important aspect to consider during AD operation is that the microbial community, specifically the methanogens, is sensitive to long-term temperature variations. Experience from large-scale operations shows that constant temperature fluctuations should not exceed ±2–3°C in order to avoid instability [131].

One quite consistent effect of operation at thermophilic instead of mesophilic temperature is a higher level of Firmicutes compared with Bacteroidetes/Proteobacteria [5, 57, 116, 118, 121, 132–136]. A high Firmicutes to Bacteroidetes ratio in mesophilic AD has been shown to correlate positively with high methane yield [137, 138]. However, an increase in this ratio has also been suggested to decrease the richness of predicted lignocellulolytic enzymes in biogas digesters, an effect attributed to lower hydrolysis in comparison with natural anaerobic systems [139]. Whether similar correlations arise in comparisons between mesophilic and thermophilic biogas communities has yet to be determined. Another characteristic feature of thermophilic communities is a higher dominance of the phylum Thermotogae [91, 114, 117, 123, 133, 135, 136, 140–143]. Members of the Thermotogae degrade polysaccharides to ethanol, acetate, CO_2, and H_2 [72, 144], but can also be involved in degradation of alcohols to CO_2 and H_2 in syntrophic association with a hydrogen-consuming partner [72, 145].

Another aspect to consider during operation at thermophilic temperature is that ammonia inhibition occurs more quickly at higher temperature, as the equilibrium between ammonium and ammonia shifts towards the latter when the temperature rises [146]. Irrespective of temperature, methanogens performing the last step in AD are among the least tolerant to ammonia and reduced methane yield, and accumulation of fatty acids is a common consequence of microbial inhibition of this group [81]. Methanogenic community changes related to temperature, often combined with increasing ammonia levels, have been reported to include positive correlations between high temperature and enhanced relative abundance of Methanobacteriales (often *Methanothermobacter*) and/or Methanomicrobiales (often *Methanoculleus*) [5, 106, 116–118, 133, 136, 140–142, 147–150]. The shift in methanogenic community often also involves a shift in acetate degradation pathway from aceticlastic methanogenesis to syntrophic acetate oxidation (SAO) [81]. However, in AD processes that seldom reach high-ammonia levels, such as AD of wastewater sludge, instant temperature changes without associated instability have been shown to be possible [143, 151, 152].

5.3 Pretreatment

Wastes rich in lignocellulose (e.g., forestry by-products, straw) or keratinase (e.g., waste from poultry, meat, and fish industries) and wastewater sludge have significant biogas potential [15, 16, 153, 154]. However, the complex floc structures of microbial cells in sewage sludge and the recalcitrant structure of lignocellulose make hydrolysis the rate-limiting step in AD systems [16, 155, 156]. Pretreatment is a well-proven approach to improve degradation of such waste. Common pretreatment strategies comprise physical (e.g., heat/pressure, irradiation, ultrasonic), chemical (e.g., acids/bases, ozonation, oxidation), and biological (addition of fungi/bacteria/enzymes under aerobic or anaerobic conditions) methods [16, 157]. The general concept of pretreatment is that it should improve the accessibility of the material to microbial degradation by disrupting the structure, changing the biomass porosity, and reducing the particle size to enhance the surface area that can be attacked. Many studies have investigated the effect on methane yield of pretreatment of various materials and many methods have shown improved process efficiency following pretreatment [16, 157, 158]. However, fewer studies have examined the influence of pretreatment on microbial communities and relationships to the increase in methane yield, and most of the studies performed to date have been on AD of waste-activated sludge, with differing results. For example, during mesophilic AD of sewage sludge, some studies have found no responses in the microbial community following thermophilic aerobic digestion or ultrasonic or alkaline pretreatment [159–161]. However, in other studies investigating mesophilic AD processes, ultrasonic, microwave, and electrokinetic pretreatments have all been shown to increase the relative abundance of *Clostridiales* (phylum Firmicutes) and Cloacimonetes and decrease the relative abundance of Proteobacteria [162, 163]. Moreover, in mesophilic AD of microalgae biomass, thermal pretreatment has been found to increase the relative abundance of the families Rikenellaceae (phylum Bacteriodetes) and Anaerolineaceae (phylum Chloroflexi) and decrease the relative abundance of the phylum Proteobacteria [164]. Using metatranscriptomic analysis, Xia et al. [165] found that low-frequency ultrasonic treatment of sludge during thermophilic digestion increased the hydrolytic activity of representatives of the phyla Bacteroidetes and Cloacimonetes and increased motility and chemotaxis in members of the phylum Thermotoga. Another noteworthy finding in that study was that, among the bacteria involved in cellulose degradation, members of the order Bacteroidales were more active than members of the Clostridiales. Both these groups contain well-known cellulose-degrading bacteria, but members of the Bacteroidales typically do not possess the cellulosomes often seen in Clostridiales. Xia et al. [165] concluded that low-frequency ultrasonic pretreatment allows enrichment of a community with high hydrolytic activity without attachment to its substrate.

For substrates other than sludge, Wang et al. [166] reported a weak effect on the microbial community structure during digestion of thermal pretreated distilled grain waste in thermophilic solid AD. Thermal and thermochemical pretreatment approaches are the most commonly used methods for lignocellulosic materials used for bioenergy production purposes [167]. Such methods are often efficient in breaking the carbohydrate polymers to soluble sugars and improving the accessibility of the substrate to microbial degradation, thus increasing the biogas yield. However, these pretreatments can also release inhibitors such as furfural, 5-HMF, vanillin, and other phenolic compounds [167]. Depending on concentration, these lignin-derived compounds have been found to be inhibitory to methanogen and to result in decreased hydrolytic activity, and major shifts have been shown to occur in both archaeal and bacterial populations (see reviews [167, 168]). However, adaptation and degradation of these compounds is possible and is suggested to involve members within the families Syntrophorhabdaceae and Synergistaceae, combined with hydrogenotrophic methanogens [167–169]. For optimized degradation of phenolic compounds, thermophilic pretreatment has been suggested [126].

The combined results from studies performed to date suggest that pretreatment mostly causes minor structural adjustments in the prevailing AD microbial community, but still impacts the activity. It is likely that the effect of pretreatment depends strongly on the prevailing operating conditions (e.g., substrate and temperature) and the activity of the microbial community. It can be anticipated that the response in microbial community structure is also linked to the physical effects of the pretreatment on the substrate. Thus, if the pretreatment enhances the solubilization of all components in the substrate, the impact on community structure will be lower than if the pretreatment increases the solubilization of one particular compound (i.e., proteins, carbohydrates, or lipids).

5.4 Loading rate and retention time

The hydraulic retention time (HRT) or solid retention time (SRT), i.e., the average time that the biomass is maintained in the digester, and the organic loading rate (OLR) are of great importance for the microbial community. A short HRT and a high OLR are often desirable in commercial biogas production plants, since they allow for high-quantity waste treatment and high biogas production (if the AD can maintain efficiency). However, SRT should exceed the microbial doubling time of prevailing microorganisms, in order to avoid washout of the consortium and thus process collapse. Immobilization of microorganisms through inclusion of support material or by allowing the formation of granular sludge, flocks, or biofilms is a strategy used in high HRT systems to support and maintain organisms with lower growth rate than the solid retention time [170, 171].

The response by the microbial community to change in OLR and HRT has been shown to vary depending on operating conditions such as temperature and composition of substrate [5, 6]. The prevailing microbial community at the time of OLR/HRT change is also important for the overall response [172]. Moreover, the feeding approach, i.e., continuous or discontinuous feeding, can be determining for community changes [173]. Changes in OLR/HRT have been shown to cause a response in most phyla dominating in AD, such as Actinobacteria, Bacteroidetes, Firmicutes, Chloroflexi, Thermotogae, Cloacimonetes, and Euryarchaeota [172–176].

In the case of increasing load, bacteria associated with hydrolytic and acidogenetic activity, such as members of the Firmicutes or Bacteroidetes, have been shown to be enriched, in parallel with accumulation of accumulation of fatty acids [172, 176–178]. Typically, acetate accumulates first and propionate accumulates if the process disturbance continues, which is assumed to be caused by limited methanogenesis and excess levels of hydrogen [5, 172, 179]. In high-solid mesophilic AD, an increase in OLR has been found to decrease the relative abundance of Firmicutes and increase that of Bacteroidetes and Candidate division WS6 [174]. During increasing OLR of protein-rich waste (blood, casein) in mesophilic AD, the order Thermoanaerobacteriales, harboring several known SAOBs (e.g., *Caldanaerobacter* and *Alkaliphilius*), has been shown to increase, while the relative abundance of *Bacillus* (Bacteroidetes) decreases [10]. In thermophilic AD of lignocellulose, decreasing the retention time from 20 to 3 days has also been shown to increase the levels of Firmicutes, while Thermotogae and Chloroflexi decrease in abundance [175]. During mesophilic AD of food waste at increasing OLR (3–7 g volatile solids L^{-1} d^{-1}) and HRT (15–20 days), a dynamic succession has been seen in different bacterial phyla (Firmicutes and Actinobacteria), while the abundance of Euryarchaeota, specifically families Methanosarcinaceae and Methanosaetaceae, increases [172].

The frequently reported increase in the genus *Methanosarcina* in response to increasing OLR has been attributed to its efficient acetate degradation capacity and robustness to stress [94]. Several studies also suggest that members of the *Methanosarcina* are important for maintained and efficient methane production under increasing OLR [172, 180]. However, members of the Methanobacteria, Methanomicrobiales, and/or Methanomassiliicoccaceae have also been observed in certain processes with a high load, depending on prevailing conditions [5, 120, 176, 179–181]. During loading by pulsed feeding, the hydrogenotrophic Methanomicrobiales have been shown to increase, favoring the consumption of propionate, most likely through hydrogen utilization. These methanogens have also been detected in high-ammonia processes operating at high OLR [179]. Ferm et al. [182] and Xu et al. [172] suggest that acetate-utilizing methanogens are critical for efficient methane production during stable performance at increasing OLR. However, with "overload" and acidification, hydrogenotrophic methanogens, such as representatives of the orders Methanomicrobiales and Methanobacteriales, become more important and dominant.

5.5 Changes in substrate composition and feeding strategies

Substrate composition is another parameter that strongly impacts the microbial community. It is well-known that co-digestion of different materials often achieves a more balanced nutrient level and improves the process performance and biogas yield

[3, 183, 184]. However, the substrate availability for a commercial biogas plant may not always be optimal and the availability can also change over time. When changing substrate composition or choosing a substrate for a new AD process, the estimated energy yield and the nutrient value of the digestate generated have to be balanced against possible problems associated with different substrates, such as ammonia inhibition, acidification, and foaming. This section reviews the microbial communities commonly observed in processes fed with protein-, carbohydrate- or fat-rich material and the microbial responses to operating challenges that often occur in these processes.

5.5.1 Protein-rich substrate

Proteins are energy-rich and contribute nutrients to the digestate, but a possible effect of ammonia inhibition has to be considered in the processing. Ammonia (NH_3) and ammonium (NH_4^+) are formed by the microbial degradation of proteins and in particularly the unionized NH_3 is toxic to microorganisms [185]. NH_3 and NH_4^+ exist in equilibrium and higher temperature and pH shift the ratio toward a higher level of ammonia. Thus, in addition to the nitrogen content, temperature and pH should be taken into account in prediction of inhibition following a change in substrate composition [186]. The aceticlastic methanogens (*Methanosaeta* sp. and certain *Methanosarcina* sp.) are considered to be most sensitive to ammonia, but if an ammonia-tolerant community is allowed to persist in the digester, the process can cope with substantially higher ammonia levels than an unadapted process [19]. An ammonia-tolerant community often includes methane formation from acetate via SAO [120, 187–193]. In SAO, acetate-oxidizing bacteria and hydrogenotrophic methanogens work in a syntrophic manner to generate methane. Bacteria species currently known to be capable of SAO belong to the genera *Thermacetogenium* [194], *Pseudothermotoga* [145, 195], *Tepidanaerobacter acetatoxydans* [196], *Clostridium* [197], and *Syntrophaceticus* [198]. Methanogenic partners in SAO are suggested to be members of the hydrogenotrophic Methanobacteriales and Methanomicrobiales (often the genus *Methanoculleus*) [81]. *Methanosarcina* is moderately ammonia-tolerant and can use both the hydrogenotrophic and aceticlastic pathways for methane formation, and can thus possibly act as a hydrogen scavenger in SAO [81, 94] or mediate the entire process, i.e., both acetate oxidation and subsequent methanogenesis [199, 200]. An increased level of protein can also affect degradation steps other than the syntrophic and methanogenic steps. For example, an increased level of protein in AD of food waste has been demonstrated to increase the abundance of the families Porphyromonadaceae, Actinomytaceae, Lactobacillaceae, and Caldicoprobacteraceae, suggesting their direct or indirect involvement in protein hydrolysis [82]. In AD of animal manure, higher protein content has been shown to increase the genera *Desulfotomaculum* and *Eubacterium* [82, 201]. High levels of ammonia have also been shown to be negatively correlated with degradation of cellulose and with some potential cellulose degraders [112].

5.5.2 Carbohydrate-rich substrate

Carbohydrate-rich materials are difficult to use in mono-digestion for biogas, since the C/N ratio becomes too high for microbial activity. Carbohydrates are thus typically co-digested with more nitrogen-rich materials. However, complex carbohydrates can pose additional challenges, such as low degradability of lignocellulosic materials, while easily accessible carbohydrates undergo fast acidogenesis that can cause acidification [202, 203]. Animal manure and sludge are commonly used in co-digestion with straw (corn, rice, tobacco, wheat) and in these processes the two orders Clostridiales (phylum Firmicutes) and Bacteroidales (phylum Bacteroidetes) often dominate. However, the phyla Proteobacter, Chloroflexi, and Fibrobacteres also often increase in response to addition of lignocellulosic materials, with some variation depending on co-digestion material and prevailing environmental conditions [118, 202, 204–208]. The microbial community structure in AD of rice straw has been shown to be influenced by temperature, with a higher ratio of Firmicutes to Bacteroidetes being reported at higher temperature [208]. In mesophilic AD of rice straw, Bacteroidetes is reported to be the most prevalent group and the abundance is not influenced by increased OLR, whereas the second most abundant Firmicutes decreases slightly [209]. Metagenomic studies have confirmed the involvement of the phyla Proteobacteria, Firmicutes, Chloroflexi, and Bacteroidetes, but also Actinomycetes, in the degradation of lignocellulose by demonstrating the existence of CAZymes (Carbohydrate-Active Enzymes) in consortia adapted to lignocellulosic materials [59, 202].

Interestingly, similar community profiles as described above are often seen in AD of material containing comparatively high levels of easily accessible carbohydrates. For example, in co-digestion of fruit and vegetable waste with pig manure, the phyla Firmicutes, Bacteroidetes, Chloroflexi, Proteobacteria, and Actinobacteria have been found to dominate, but the numbers of Firmicutes decrease when the fraction of fruit and vegetable waste (with the highest levels of carbohydrates) decreases [210]. In mesophilic AD of potato and cabbage waste (alone or in combination), members of the phyla Spirochaete, Bacteroidetes, Firmicutes, and Proteobacteria vary in numbers depending on the substrate combination [203]. In a study examining addition of cellulose and xylan to wastewater sludge, it was found that this increased the relative abundance of the bacterial genus *Clostridium* (phylum Firmicutes), whereas the levels of the bacterial phyla Thermotogae and Bacteroidetes decreased [211]. In thermophilic AD of cattle manure involving addition of easily degraded carbohydrates in the form of glucose, the genus *Lactobacillus* (class Bacilli) has been shown to increase [201]. The methanogenic communities identified in various studies on carbohydrate-rich material show diverging structures and appear to be primarily shaped by the co-substrate and prevailing environmental conditions. For example, during straw co-digestion with cow manure or digestion of straw alone, *Methanosarcina* or *Methanosaeta* often dominate [204–207, 208, 209]. However, with increasing nitrogen level, temperature, OLR, and/or carbohydrate accessibility, the contribution of hydogenotropic methanogenesis increases, involving *Methanoculleus, Methanothermobacter*, and *Methanobacterium* [201–203, 208, 209].

5.5.3 Lipid-rich substrate

Lipids are energy-rich and different fat-rich substrates are often used to boost biogas production from sewage and manure [212–215]. Degradation of fat results in glycerol and LCFA, with the latter being a known microbial inhibitor [23]. The bacteria *Syntrophomonas* (family Syntrophomonadaceae) is commonly enriched in mesophilic co-digestion of lipid-rich materials [216–224] and even represents as much as 30–40% of the total bacterial community during degradation of LCFA [218, 225]. Moreover, it has been reported [218] that pulse feeding of oleate, instead of continuous feeding of oleate, increases the conversion rates of oleate and acetate and induces greater metabolic flexibility within the LCFA-degrading community dominated by *Syntrophomonas* population [76]. In thermophilic degradation of animal manure, addition of oleate has been shown to increase the relative abundance of the glycerol- and inositol-fermenting *Megamonas* (phylum Firmicutes) [201], whereas in mesophilic AD increased levels of glycerol/glycerin enrich the phyla Cloacamonas [226] and Thermotogae in AD of wastewater sludge [227] and the genus *Trichococcus* and family Syntrophomonadaceae in AD of brewery wastewater [228]. *Methanoculleus*, *Methanobacterium*, and *Methanospirillum* have been proposed as important hydrogen-utilizing partners for LCFA-degrading bacteria, whereas *Methanosarcina* has been suggested to act both as a hydrogen and acetate consumer [216, 229, 230]. However, in pulse feeding of oleate, *Methanosaeta* increases in importance relative to *Methanosarcina*, along with higher abundance of *Methanoculleus* compared with *Methanobacterium*. This was suggested by the authors to be a consequence of higher acetate affinity and tolerance for LCFA by *Methanosaeta* and higher affinity for hydrogen by *Methanoculleus* [218]. In another study, an increased level of the hydrogenotrophic *Methanoculleus* and *Methanobrevibacter* was linked to increased methane production from oleate, driven by enhanced concentration of sulfide [224]. In addition to aceticlastic methanogensis, acetate degradation has also been shown to proceed via syntrophic acetate oxidation during LCFA conversion, which is likely linked to high-ammonia level [216].

5.6 Addition of trace elements

Trace element deficiency can severely limit microbial activity and cause accumulation of fatty acids, process instability and decreased methane yield from food waste [21, 120, 231, 232], slaughterhouse waste [233, 234], crop material [235], stillage [236], and animal manure, when used as a single substrate or as co-substrate [237, 238]. In this regard, it is important to consider the level of sulfide, which is primarily formed through protein degradation. Sulfide forms complexes with metals, which decreases the bioavailability of trace elements essential for microbial activity [239–241]. In addition, temperature has been suggested to impact nutrient bioavailability and nutrient requirements [242, 243]. However, the actual impact of different temperatures on the availability of trace metals has yet to be established.

The trace elements such as cobalt, nickel, iron, molybdenum, and tungsten are essential trace elements, especially for acetogenic and methanogenic microorganisms [244–246]. So far, mainly methanogenic abundance has been shown to be influenced by trace element addition in AD, while less is known about the response in the bacterial community. Thus, it is not clear whether the improved degradation of LCFA and VFA with trace element addition is caused solely by improved activity of methanogens or also improved activity of the syntrophic community. Trace elements have demonstrated to have a pronounced effect on the methanogenic community, including increased abundance or predicted stimulatory effects on the genus *Methanoculleus* [120, 247] and increased abundance of the order Methanosarcinales [200] and the genus *Methanobrevibacter* (order Methanobacteriales), all in mesophilic AD [247]. *Methanoculleus* has also been proposed to have a more efficient strategy than *Methanosarcina* for stabilizing its energy balance, and thus can cope more successfully with trace element limitation [248, 249]. Interestingly, despite improved VFA conversion following trace element addition, SAO-dominated AD processes are reported to show no or decreased abundance of the known syntrophic acetate oxidizers *S. schinkii*, *T. acetatoxydanse*, and *C. ultunense* [120, 200].

5.7 Bioaugmentation

The approach of adding microorganisms to the anaerobic process is based on the belief that slow degradation is due to the absence or low abundance of efficient populations responsible for the particular degradation step. Bioaugmentation could thus shorten the time of microbial adaptation to certain environmental conditions/inhibitors and/or improve methane yield from specific substrates. Since the hydrolytic and methanogenic steps generally appear to be bottlenecks in AD systems, bioaugmentation efforts to date have most commonly been directed at enhancing these two steps. However, bioaugmentation has also been evaluated for improving the transition to psychrophilic temperature, to overcome inhibition of ammonia and reduce the time following overload [250].

For the degradation of lignocellulosic material in the biogas process, bioaugmentation with cellulose-degrading bacteria, hydrolytic enzymes, and anaerobic fungi has been suggested as a promising method to increase methane production from lignocellulosic materials [251–254]. Microorganisms that have so far shown positive results on methane yield include the cellulolytic bacteria *Clostridium cellulolyticulm*, *Acetobacteroides hydrogenigenes*, and *Caldicellulosiruptor lactoaceticus* (*Caldicellulosiruptor*) and the fungus *Piromyces rhizinflata*. A mix of cultures of different *Clostridium* sp. and different hemicellulose and cellulolytic bacteria has also been shown to produce positive results [250], while a mixed consortium with high endoglucanase activity has been found to result in increased biogas production from maize silage [255]. For addition of enzymes, investigations have shown mixed effects, ranging from no effect at all on rate or yield, to increased biogas yield only, or increased rate only (summarized in [252]). A likely explanation for the nonconclusive results from addition of enzyme/organisms is differences in the environmental conditions prevailing in the digester, such pH and ammonia level, which vary greatly depending on substrate. For example, a clear correlation between inefficient cellulose degradation and high-ammonia levels has been demonstrated [53]. The amount of added microorganisms has also been suggested to be of critical importance [250]. For enzyme addition, another possible reason behind the variation in results is that the hydrolytic enzymes investigated so far have mainly originated from nonbiogas environments and have a very short activity lifetime (<24 h) in the biogas process, which restricts the hydrolytic activity within these systems [256]. However, a study investigating the effects of addition of enzymes or microbes retrieved from a specific biogas environment has found promising results [252]. In that study, these enzymes were found to be active and stable in the environment and had a profound effect on both the biogas production rate and yield from forage ley [252]. Moreover, Azman et al. [257] have demonstrated that addition of hydrolytic enzymes to a cellulose and xylan-fed digester operating at 30°C can counteract the inhibitory effects of humic acid on hydrolysis efficiency.

The degradation of fats has been shown to be stimulated by the addition of hydrolyzing enzymes (lipases) or fat-degrading bacteria (*Syntrophomonas zehnderi* and *Clostridium lundense*) [250, 258], whereas addition of a co-culture of *Syntrophomonas zehnderi* and *Methanobacterium formicicum* is reported to have no effect in AD of fat-rich wastewater [259]. For protein, bioaugmentation with *Coprothermobacter proteolyticus* has been shown to improve hydrolysis and fermentation in waste-activated sludge [260]. Another factor to consider when attempting to improve the degradation of fat and protein is increased release of LCFA and ammonia. For example, high concentrations of lipases have been shown to inhibit the process, probably due to the release of LCFA. Moreover, LCFA and ammonia have been shown to have additive effects, so that the process becomes more severely inhibited if both are present at relatively high concentrations [205].

Previous attempts to increase the stability and activity of the methanogenic community have included addition of *Methanosarcina* sp. during start-up [111]. Moreover, bioaugmentation with syntrophic-acetate degrading co-cultures and with ammonia-tolerant *Methanoculleus bourgensis* has been tested with the aim of preventing ammonia inhibition of the process [189, 261, 262]. Test results in that case revealed that addition of syntrophic co-cultures did not facilitate a dynamic transition from aceticlastic methanogenesis to SAO, whereas addition of ammonia-tolerant *M. bourgensis* improved adaptation to gradually increased ammonia concentrations under mesophilic conditions.

6. Conclusions

Biogas production through anaerobic digestion enables recovery of renewable energy and of nutrients from various organic waste materials and is thus highly important for the transition to a more sustainable society. The performance and stability of the biodigestion process is highly dependent on an array of different microbial groups, and their networks and functions are in turn influenced by substrate characteristic and operating parameters. With recent advances in molecular techniques, knowledge about anaerobic microorganisms and their response to various operating conditions has increased tremendously. This knowledge has enabled the development of more controlled management and monitoring approaches, to ensure high process efficiency and stability. However, with increasing knowledge about the microbiology of biogas processes, it has also become evident that the microbiota involved is even more complicated and difficult to visualize than initially thought, particularly as many members within a particular genus are often able to degrade chemically very different compounds. Moreover, many organisms belong to candidate phyla or are even unknown, and remain to be isolated and characterized for full understanding of their role in the biogas system. Thus, in order to establish effective operating policies to achieve maximum biogas process performance, it is important to improve understanding about microorganisms and their functions and to further develop a predictive understanding of the interplay between microbial community structure and operating parameters and performance.

Conflict of interest

None declared.

Author details

Maria Westerholm and Anna Schnürer*
Department of Molecular Sciences, Swedish University of Agricultural Sciences, Uppsala, Sweden

*Address all correspondence to: anna.schnurer@slu.se

References

[1] Weiland P. Biogas production: Current state and perspective. Applied Microbiology and Biotechnology. 2010;**85**:849-860. DOI: 10.1007/s00253-009-2246-7

[2] Risberg K, Cederlund H, Pell M, Arthurson V, Schnturer A. Comparative characterization of digestate versus pig slurry and cow manure—Chemical composition and effects on soil microbial activity. Waste Management. 2017;**61**:529-538. DOI: 10.1016/j.wasman.2016.12.016

[3] Mata-Alvarez J, Dosta J, Romero-Güiza MS, Fonoll X, Peces M, Astals S. A critical review on anaerobic co-digestion achievements between 2010 and 2013. Renewable and Sustainable Energy Reviews. 2014;**36**:412-427. DOI: 10.1016/j.rser.2014.04.039

[4] Romero-Gueiza MS, Vila JJ, Mata-Alvarez J, Chimenos JM, Astals S. The role of additives on anaerobic digestion: A review. Renewable & Sustainable Energy Reviews. 2016;**58**:1486-1499. DOI: 10.1016/j.rser.2015.12.094

[5] Westerholm M, Isaksson S, Karlsson Lindsjö O, Schnürer A. Microbial community adaptability to altered temperature conditions determines the potential for process optimisation in biogas production. Applied Energy. 2018;**226**:838-848. DOI: 10.1016/j.apenergy.2018.06.045

[6] Ferguson RMW, Coulon F, Villa R. Organic loading rate: A promising microbial management tool in anaerobic digestion. Water Research. 2016;**100**:348-356. DOI: 10.1016/j.watres.2016.05.009

[7] Lindmark J, Thorin E, Fdhila RB, Dahlquist E. Effects of mixing on the result of anaerobic digestion: Review. Renewable and Sustainable Energy Reviews. 2014;**40**:1030-1047. DOI: 10.1016/j.rser.2014.07.182

[8] Schnürer A, Bohn I, Moestedt J. Protocol for start-up and operation of CSTR biogas processes. In: TJea MG, editor. Hydrocarbon and Lipid Microbiology Protocols. Springer Protocols Handbooks. Berlin Heidelberg: Springer-Verlag; 2017. pp. 171-200

[9] Robles G, Nair RB, Kleinsteuber S, Nikolausz M, Horvath IS. Biogas production: Microbiological aspects. In: Tabatabaei M, Ghanavati H, editors. Biogas. Biofuel and Biorefinery Technologies. Cham: Springer; 2018. pp. 163-198. DOI: 10.1007/978-3-319-77335-3_7

[10] Kovács E, Wirth R, Mároti G, Bagi Z, Rákhely G, Kovács KL. Biogas production from protein-rich biomass: Fed-batch anaerobic fermentation of casein and of pig blood and associated changes in microbial community composition. PLoS One. 2013;**8**:1-18. DOI: 10.1371/journal.pone.0077265

[11] He Q, Li L, Zhao X, Qu L, Wu D, Peng X. Investigation of foaming causes in three mesophilic food waste digesters: Reactor performance and microbial analysis. Science Reporter. 2017;**7**:13701. DOI: 10.1038/s41598-017-14258-3

[12] Lienen T, Kleyboecker A, Verstraete W, Wuerdemann H. Foam formation in a downstream digester of a cascade running full-scale biogas plant: Influence of fat, oil and grease addition and abundance of the filamentous bacterium *Microthrix parvicella*. Bioresource Technology. 2014;**153**:1-7. DOI: 10.1016/j.biortech.2013.11.017

[13] Sánchez ÒJ, Cardona CA. Trends in biotechnological production of fuel ethanol from different feedstocks. Bioresource Technology. 2008;**99**:5270-5295. DOI: 10.1016/j.biortech.2007.11.013

[14] Azman S, Khadem AF, van Lier JB, Zeeman G, Plugge CM. Presence and

role of anaerobic hydrolytic microbes in conversion of lignocellulosic biomass for biogas production. Critical Reviews in Environmental Science and Technology. 2015;**45**:2523-2564. DOI: 10.1080/10643389.2015.1053727

[15] Lynd LR, Weimer PJ, van Zyl WH, Pretorius IS. Microbial cellulose utilization: Fundamentals and biotechnology. Microbiology and Molecular Biology Reviews. 2002;**66**:506-577. DOI: 10.1128/MMBR.66.3.506-577.2002

[16] Wagner AO, Lackner N, Mutschlechner M, Prem EM, Markt R, Illmer P. Biological pretreatment strategies for second-generation lignocellulosic resources to enhance biogas production. Energies. 2018;**11**:1797. DOI: 10.3390/en11071797

[17] Moestedt J, Nilsson Påledal S, Schnürer A, Nordell E. Biogas production from thin stillage on an industrial scale—Experience and optimisation. Energies. 2013;**6**: 5642-5655. DOI: 10.3390/en6115642

[18] Solli L, Elise Håvelsrud O, Horn SJ, Gunn Rike A. A metagenomic study of the microbial communities in four parallel biogas reactors. Biotechnology for Biofuels. 2014;**7**:1-15. DOI: 10.1186/s13068-014-0146-2

[19] Rajagopal R, Massé DI, Singh G. A critical review on inhibition of anaerobic digestion process by excess ammonia. Bioresource Technology. 2013;**143**:632-641. DOI: 10.1016/j.biortech.2013.06.030

[20] Luo G, Fotidis IA, Angelidaki I. Comparative analysis of taxonomic, functional, and metabolic patterns of microbiomes from 14 full-scale biogas reactors by metagenomic sequencing and radioisotopic analysis. Biotechnology for Biofuels. 2016;**9**:1-12. DOI: 10.1186/s13068-016-0465-6

[21] Banks CJ, Zhang Y, Jiang Y, Heaven S. Trace elements requirements for stable food waste digestion at elevated ammonia concentrations. Bioresource Technology. 2012;**104**:127-135. DOI: 10.1016/j.biortech.2011.10.068

[22] Angelidaki I, Karakashev D, Batstone DJ, Plugge CM, Stams AJM. Biomethane and its potential. Methods in Enzymology. 2011;**494**:327-351. DOI: 10.1016/B978-0-12-385112-3.00016-0

[23] Sousa DZ, Smidt H, Alves MM, Stams AJM. Ecophysiology of syntrophic communities that degrade saturated and unsaturated long-chain fatty acids. FEMS Microbiology Ecology. 2009;**68**:257-272. DOI: 10.1111/j.1574-6941.2009.00680.x

[24] Luostarinen S, Luste S, Sillanpää M. Increased biogas production at wastewater treatment plants through co-digestion of sewage sludge with grease trap sludge from a meat processing plant. Bioresource Technology. 2009;**100**:79-85. DOI: 10.1016/j.biortech.2008.06.029

[25] Affes M, Aloui F, Hadrich F, Loukil S, Sayadi S. Effect of bacterial lipase on anaerobic co-digestion of slaughterhouse wastewater and grease in batch condition and continuous fixed-bed reactor. Lipids in Health and Disease. 2017;**16**:195. DOI: 10.1186/s12944-017-0587-2

[26] Hanaki K. Mechanism of inhibition caused by long-chain fatty acids in anaerobic digestion process. Biotechnology and Bioengineering. 1981;**23**:1591-1610

[27] Cuetos MJ, Gómez X, Otero M, Morán A. Anaerobic digestion of solid slaughterhouse waste (SHW) at laboratory scale: Influence of co-digestion with the organic fraction of municipal solid waste (OFMSW). Biochemical Engineering Journal. 2008;**40**:99-106. DOI: 10.1016/j.bej.2007.11.019

[28] Cirne DG, Lehtomäki A, Björnsson L, Blackall LL. Hydrolysis and microbial community analyses in two-stage anaerobic digestion of energy crops. Journal of Applied Microbiology. 2007;**103**:516-527. DOI: 10.1111/j.1365-2672.2006.03270.x

[29] Kazda M, Langer S, Bengelsdorf FR. Fungi open new possibilities for anaerobic fermentation of organic residues. Energy, Sustainability and Society. 2014;**4**:6. DOI: 10.1186/2192-0567-4-6

[30] Schnürer A. Biogas production: Microbiology and technology. Advances in Biochemical Engineering/ Biotechnology. In: Hatti-Kaul R, Mamo G, Mattiasson B, editors. Switzerland: Springer International Publishing; 2016;**156**:195-234. DOI: 10.1007/10_2016_5

[31] Tsavkelova EA, Netrusov AI. Biogas production from cellulose containing substrates: A review. Applied Biochemistry and Microbiology. 2012;**48**:421-433. DOI: 10.1134/ S0003683812050134

[32] Ding S, Liu Y, Zeng Y, Himmel ME, Baker JO, Bayer EA. How does plant cell wall nanoscale architecture correlate with enzymatic digestibility? Science. 2012;**338**:1055-1060. DOI: 10.1126/ science.1227491

[33] Ramsay IR, Pullammanappallil PC. Protein degradation during anaerobic wastewater treatment: Derivation of stoichiometry. Biodegradation. 2001;**12**:247-257. DOI: 10.1023/A:1013116728817

[34] Schink B, Stams AJM. Syntrophism among prokaryotes. In: Dworkin M, Falkow S, Rosenberg E, Schleifer KH, Stackebrandt E, editors. The Prokaryotes. New York: Springer-Verlag; 2006. pp. 309-335

[35] Mendes AA, Oliveira PC, de Castro HF. Properties and biotechnological applications of porcine pancreatic lipase. Journal of Molecular Catalysis B: Enzymatic. 2012;**78**:119-134. DOI: 10.1016/j.molcatb.2012.03.004

[36] Stergiou P, Foukis A, Filippou M, Koukouritaki M, Parapouli M, Theodorou LG, et al. Advances in lipase-catalyzed esterification reactions. Biotechnology Advances. 2013;**31**:1846-1859. DOI: 10.1016/j. biotechadv.2013.08.006

[37] Weng C, Jeris JS. Biochemical mechanisms in the methane fermentation of glutamic and oleic acids. Water Research. 1974;**10**:9-18

[38] Ma J, Zhao Q-B, Laurens LLM, Jarvis EE, Nagle NJ, Chen S, et al. Mechanism, kinetics and microbiology of inhibition caused by long-chain fatty acids in anaerobic digestion of algal biomass. Biotechnology for Biofuels. 2015;**8**:141. DOI: 10.1186/s13068-015-0322-z

[39] Kougias PG, Angelidaki I. Biogas and its opportunities—A review. Frontiers of Environmental Science & Engineering. 2018;**12**:14. DOI: 10.1007/ s11783-018-1037-8

[40] Kandylis P, Bekatorou A, Pissaridi K, Lappa K, Dima A, Kanellaki M, et al. Acidogenesis of cellulosic hydrolysates for new generation biofuels. Biomass and Bioenergy. 2016;**91**:210-216. DOI: 10.1016/j.biombioe.2016.05.006

[41] Ragsdale S, Pierce E. Acetogenesis and the Wood-Ljungdahl pathway of CO_2 fixation. Biochimica et Biophysica Acta. 2008;**1784**:1873-1898. DOI: 10.1016/j.bbapap.2008.08.012

[42] Drake HL, Gossner AS, Daniel SL. Old acetogens, new light. Annals of the New York Academy of Sciences. 2008;**1125**:100-128. DOI: 10.1196/ annals.1419.016

[43] Worm P, Müller N, Plugge CM, Stams AJM, Schink B. Syntrophy

in methanogenic degradation. In: Hackstein JHP, editor. (Endo)symbiotic Methanogenic Archaea. Berlin: Springer-Verlag; 2010. pp. 143-173

[44] Schink B. Energetics of syntrophic cooperation in methanogenic degradation. Microbiology and Molecular Biology Reviews. 1997;**61**:262-280

[45] Dong X, Plugge CM, Stams AJM. Anaerobic degradation of propionate by a mesophilic acetogenic bacterium in coculture and triculture with different methanogens. Applied and Environmental Microbiology. 1994;**60**:2834-2838

[46] Liu F, Whitman WB. Metabolic, phylogenetic, and ecological diversity of the methanogenic archaea. Annals of the New York Academy of Sciences. 2008;**1125**:171-189. DOI: 10.1196/annals.1419.019

[47] Costa KC, Leigh JA. Metabolic versatility in methanogens. Current Opinion in Biotechnology. 2014;**29**:70-75. DOI: 10.1016/j.copbio.2014.02.012

[48] Lebuhn M, Hanreich A, Klocke M, Schlüter A, Bauer C, Pérez CM. Towards molecular biomarkers for biogas production from lignocellulose-rich substrates. Anaerobe. 2014;**29**:10-21. DOI: 10.1016/j.anaerobe.2014.04.006

[49] Krause L, Diaz NN, Edwards RA, Gartemann K, Krömeke H, Neuweger H, et al. Taxonomic composition and gene content of a methane-producing microbial community isolated from a biogas reactor. Journal of Biotechnology. 2008;**136**:91-101. DOI: 10.1016/j.jbiotec.2008.06.003

[50] Klocke M, Mähnert P, Mundt K, Souidi K, Linke B. Microbial community analysis of a biogas-producing completely stirred tank reactor fed continuously with fodder beet silage as mono-substrate. Systematic and Applied Microbiology. 2007;**30**:139-151. DOI: 10.1016/j.syapm.2006.03.007

[51] Wang H, Lehtomäki A, Tolvanen K, Puhakka J, Rintala J. Impact of crop species on bacterial community structure during anaerobic co-digestion of crops and cow manure. Bioresource Technology. 2009;**100**:2311-2315. DOI: 10.1016/j.biortech.2008.10.040

[52] Liu T, Sun L, Müller B, Schnürer A. Importance of inoculum source and initial community structure for biogas production from agricultural substrates. Bioresource Technology. 2017;**245**:768-777. DOI: 10.1016/j.biortech.2017.08.213

[53] Sun L, Liu T, Müller B, Schnürer A. The microbial community structure in industrial biogas plants influences the degradation rate of straw and cellulose in batch tests. Biotechnology for Biofuels. 2016;**9**:128. DOI: 10.1186/s13068-016-0543-9

[54] Jia Y, Ng S, Lu H, Cai M, Lee PKH. Genome-centric metatranscriptomes and ecological roles of the active microbial populations during cellulosic biomass anaerobic digestion. Biotechnology for Biofuels. 2018;**11**:117. DOI: 10.1186/s13068-018-1121-0

[55] Campanaro S, Treu L, Kougias PG, Francisci DD, Valle G, Angelidaki I. Metagenomic analysis and functional characterization of the biogas microbiome using high throughput shotgun sequencing and a novel binning strategy. Biotechnology for Biofuels. 2016;**9**:26. DOI: 10.1186/s13068-016-0441-1

[56] Koeck DE, Wibberg D, Maus I, Winkler A, Albersmeier A, Zverlov VV, et al. Complete genome sequence of the cellulolytic thermophile *Ruminoclostridium cellulosi* wild-type strain DG5 isolated from a thermophilic biogas plant. Journal of Biotechnology. 2014;**188**:136-137. DOI: 10.1016/j.jbiotec.2014.08.024

[57] Liu CM, Wachemo AC, Tong H, Shi SH, Zhang L, Yuan HR, et al. Biogas production and microbial community

properties during anaerobic digestion of corn stover at different temperatures. Bioresource Technology. 2018;**261**:93-103. DOI: 10.1016/j.biortech.2017.12.076

[58] Limam RD, Chouari R, Mazéas L, Wu T, Li T, Grossin-Debattista J, et al. Members of the uncultured bacterial candidate division WWE1 are implicated in anaerobic digestion of cellulose. Microbiology Open. 2014;**3**:157-167. DOI: 10.1002/mbo3.144

[59] Wang C, Dong D, Wang H, Mueller K, Qin Y, Wang H, et al. Metagenomic analysis of microbial consortia enriched from compost: New insights into the role of Actinobacteria in lignocellulose decomposition. Biotechnology for Biofuels. 2016;**9**:22. DOI: 10.1186/s13068-016-0440-2

[60] Dollhofer V, Podmirseg SM, Callaghan TM, Griffith GW, Fliegerova K. Anaerobic fungi and their potential for biogas production. In: Guebitz G, Bauer A, Bochmann G, Gronauer A, Weiss S, editors. Biogas Science and Technology. Advances in Biochemical Engineering/Biotechnology. Vol. 151. Cham: Springer; 2015. DOI: 10.1007/978-3-319-21993-6_2

[61] Cheng YF, Shi QC, Sun RL, Liang D, Li YF, Li YQ, et al. The biotechnological potential of anaerobic fungi on fiber degradation and methane production. World Journal of Microbiology and Biotechnology. 2018;**34**:155. DOI: 10.1007/s11274-018-2539-z

[62] Baena S, Fardeau ML, Ollivier B, Labat M, Thomas P, Garcia JL, et al. *Aminomonas paucivorans* gen. nov., sp. nov., a mesophilic, anaerobic, amino acid-utilizing bacterium. International Journal of Systematic Bacteriology. 1999;**49**:975-982. DOI: 10.1099/00207713-49-3-975

[63] Baena S, Fardeau M, Woo THS, Ollivier B, Labat M, Patel BKC. Phylogenetic relationships of three amino-acid-utilizing anaerobes, *Selenomonas acidaminovorans*, '*Selenomonas acidaminophila*' and *Eubacterium acidaminophilum*, as inferred from partial 16S rDNA nucleotide sequences and proposal of *Thermanaerovibrio acidaminovorans* gen. nov., comb. nov. and *Anaeromusa acidaminophila* gen. nov., comb. nov. International Journal of Systematic Bacteriology. 1999;**49**:969-974

[64] Baena S, Fardeau ML, Labat M, Ollivier B, Garcia JL, Patel BKC. *Aminobacterium mobile* sp. nov., a new anaerobic amino acid-degrading bacterium. International Journal of Systematic and Evolutionary Microbiology. 2000;**50**:259-264

[65] Ueki A, Abe K, Suzuki D, Kaku N, Watanabe K, Ueki K. *Anaerosphaera aminiphila* gen. nov., sp nov., a glutamate-degrading, Gram-positive anaerobic coccus isolated from a methanogenic reactor treating cattle waste. International Journal of Systematic and Evolutionary Microbiology. 2009;**59**:3161-3167. DOI: 10.1099/ijs.0.011858-0

[66] Plugge CM, Balk M, Zoetendal EG, Stams AJM. *Gelria glutamica* gen. nov., sp nov., a thermophilic, obligately syntrophic, glutamate-degrading anaerobe. International Journal of Systematic and Evolutionary Microbiology. 2002;**52**:401-407. DOI: 10.1099/ijs.0.01949-0

[67] Tomazetto G, Hahnke S, Maus I, Wibberg D, Puehler A, Schlueter A, et al. Complete genome sequence of *Peptoniphilus* sp strain ING2-D1G isolated from a mesophilic lab-scale completely stirred tank reactor utilizing maize silage in co-digestion with pig and cattle manure for biomethanation. Journal of Biotechnology. 2014;**192**:59-61. DOI: 10.1016/j.jbiotec.2014.09.011

[68] Tang Y, Shigematsu T, Morimura S, Kida K. Microbial community analysis of mesophilic anaerobic protein degradation

process using bovine serum albumin (BSA)-fed continuous cultivation. Journal of Bioscience and Bioengineering. 2005;**99**:150-164. DOI: 10.1263/jbb.99.150

[69] Hahnke S, Langer T, Klocke M. *Proteiniborus indolifex* sp nov., isolated from a thermophilic industrial-scale biogas plant. International Journal of Systematic and Evolutionary Microbiology. 2018;**68**:824-828. DOI: 10.1099/ijsem.0.002591

[70] Hernandez-Eugenio G, Fardeau ML, Cayol JL, Patel BK, Thomas P, Macarie H, et al. *Sporanaerobacter acetigenes* gen. nov., sp. nov., a novel acetogenic, facultatively sulfur-reducing bacterium. International Journal of Systematic and Evolutionary Microbiology. 2002;**52**:1217-1223. DOI: 10.1099/00207713-52-4-1217

[71] Hahnke S, Langer T, Koeck DE, Klocke M. Description of *Proteiniphilum saccharofermentans* sp. nov., *Petrimonas mucosa* sp. nov. and *Fermentimonas caenicola* gen. nov., sp. nov., isolated from mesophilic laboratory-scale biogas reactors, and emended description of the genus *Proteiniphilum*. International Journal of Systematic and Evolutionary Microbiology. 2016;**66**:1466-1475. DOI: 10.1099/ijsem.0.000902

[72] Stolze Y, Bremges A, Maus I, Pühler A, Sczyrba A, Schlüter A. Targeted in situ metatranscriptomics for selected taxa from mesophilic and thermophilic biogas plants. Microbial Biotechnology. 2017;**11**:667-679. DOI: 10.1111/1751-7915.12982

[73] Li A, Chu Y, Wang X, Ren L, Yu J, Liu X, et al. A pyrosequencing-based metagenomic study of methane-producing microbial community in solid-state biogas reactor. Biotechnology for Biofuels. 2013;**6**:3. DOI: 10.1186/1754-6834-6-3

[74] Petropoulos E, Dolfing J, Yu Y, Wade MJ, Bowen EJ, Davenporta RJ, Curtis TP. Lipolysis of domestic wastewater in anaerobic reactors operating at low temperatures. Environmental Science: Water Research & Technology. 2018;**4**:1002. DOI: 10.1039/c8ew00156a

[75] St-Pierre B, Wright A-DG. Comparative metagenomic analysis of bacterial populations in three full-scale mesophilic anaerobic manure digesters. Applied Microbiology and Biotechnology. 2014;**98**:2709-2717. DOI: 10.1007/s00253-013-5220-3

[76] Ziels RM, Sousa DZ, Stensel HD, Beck DAC. DNA-SIP based genome-centric metagenomics identifies key long-chain fatty acid-degrading populations in anaerobic digesters with different feeding frequencies. The ISME Journal. 2018;**12**:112-123. DOI: 10.1038/ismej.2017.143

[77] Worm P, Koehorst JJ, Visser M, Sedano-Núnez VT, Schaap PJ, Plugge CM, et al. A genomic view on syntrophic versus non-syntrophic lifestyle in anaerobic acid degrading communities. Biochimica et Biophysica Acta. 2014;**1837**:2004-2016. DOI: 10.1016/j.bbabio.2014.06.005

[78] Nobu MK, Narihiro T, Rinke C, Kamagata Y, Tringe SG, Woyke T, et al. Microbial dark matter ecogenomics reveals complex synergistic network in a methanogenic bioreactor. The ISME Journal. 2015;**9**:1710-1722. DOI: 10.1038/ismej.2014.256

[79] Pelletier E, Kreimeyer A, Bocs S, Rouy Z, Gyapay G, Chouari R, et al. "*Candidatus* Cloacamonas Acidaminovorans": Genome sequence reconstruction provides a first glimpse of a new bacterial division. Journal of Bacteriology. 2008;**190**:2572-2579. DOI: 10.1128/JB.01248-07

[80] Li J, Rui J, Yao M, Zhang S, Yan X, Wang Y, et al. Substrate type and free ammonia determine bacterial community structure in full-scale

mesophilic anaerobic digesters treating cattle or swine manure. Frontiers in Microbiology. 2015;**6**:1337. DOI: 10.3389/fmicb.2015.01337

[81] Westerholm M, Moestedt J, Schnürer A. Biogas production through syntrophic acetate oxidation and deliberate operating strategies for improved digester performance. Applied Energy. 2016;**179**:124-135. DOI: 10.1016/j.apenergy.2016.06.061

[82] Müller B, Sun L, Westerholm M, Schnürer A. Bacterial community composition and *fhs* profiles of low- and high-ammonia biogas digesters reveal novel syntrophic acetate-oxidising bacteria. Biotechnology for Biofuels. 2016;**9**:1-18. DOI: 10.1186/s13068-016-0454-9

[83] Zakrzewski M, Goesmann A, Jaenicke S, Jünemann S, Eikmeyer F, Szczepanowski R, et al. Profiling of the metabolically active community from a production-scale biogas plant by means of high-throughput metatranscriptome sequencing. Journal of Biotechnology. 2012;**158**:248-258. DOI: 10.1016/j.jbiotec.2012.01.020

[84] Westerholm M, Müller B, Singh A, Karlsson Lindsjö O, Schnürer A. Detection of novel syntrophic acetate-oxidising bacteria from biogas processes by continuous acetate enrichment approaches. Microbial Biotechnology. 2018;**11**:680-693. DOI: 10.1111/1751-7915.13035

[85] Mosbaek FHK, Mulat DG, Albertsen M, Ward AJ, Feilberg A, et al. Identification of syntrophic acetate-oxidizing bacteria in anaerobic digesters by combined protein-based stable isotope probing and metagenomics. The ISME Journal. 2016:1-14. DOI: 10.1038/ismej.2016.39

[86] Bassani I, Kougias PG, Treu L, Angelidaki I. Biogas upgrading via hydrogenotrophic methanogenesis in two-stage continuous stirred tank reactors at mesophilic and thermophilic conditions. Environmental Science & Technology. 2015;**49**:12585-12593. DOI: 10.1021/acs.est.5b03451

[87] Ito T, Yoshiguchi K, Ariesyady HD, Okabe S. Identification of a novel acetate-utilizing bacterium belonging to Synergistes group 4 in anaerobic digester sludge. The ISME Journal. 2011;**5**: 1844-1856. DOI: 10.1038/ismej.2011.59

[88] Ho D, Jensen P, Batstone D. Effects of temperature and hydraulic retention time on acetotrophic pathways and performance in high-rate sludge digestion. Environmental Science & Technology. 2014;**48**:6468-6476. DOI: 10.1021/es500074j

[89] Lee S, Park J, Kim SH, Yu BJ, Yoon J, Park H. Evidence off syntrophic acetate oxidation by *Spirochaetes* during anaerobic methane production. Bioresource Technology. 2015;**190**:543-549. DOI: 10.1016/j.biortech.2015.02.066

[90] Ruiz-Sánchez J, Campanaro S, Guivernau M, Fernández B, Prenafeta-Boldú FX. Effect of ammonia on the active microbiome and metagenome from stable full-scale digesters. Bioresource Technology. 2018;**250**:513-522. DOI: 10.1016/j.biortech.2017.11.068

[91] Maus I, Koeck DE, Cibis KG, Hahnke S, Kim YS, Langer T, et al. Unraveling the microbiome of a thermophilic biogas plant by metagenome and metatranscriptome analysis complemented by characterization of bacterial and archaeal isolates. Biotechnology for Biofuels. 2016;**9**:171. DOI: 10.1186/s13068-016-0581-3

[92] Heyer R, Kohrs F, Benndorf D, Rapp E, Kausmann R, Heiermann M, et al. Metaproteome analysis of the microbial communities in agricultural

biogas plants. New Biotechnology. 2013;**30**:614-622. DOI: 10.1016/j.nbt.2013.01.002

[93] Enzmann F, Mayer F, Rother M, Holtmann D. Methanogens: Biochemical background and biotechnological applications. AMB Express. 2018;**8**:1. DOI: 10.1186/s13568-017-0531-x

[94] De Vrieze J, Hennebel T, Boon N, Verstraete W. *Methanosarcina*: The rediscovered methanogen for heavy duty biomethanation. Bioresource Technology. 2012;**112**:1-9. DOI: 10.1016/j.biortech.2012.02.079

[95] Nubo KM, Narihiro T, Kuroda K, Mei R, Liu WT. Chasing the elusive Euryarchaeota class WSA2: Genomes reveal a uniquely fastidious methyl-reducing methanogen. The ISME Journal. 2016;**10**:2478-2487. DOI: 1751-7362/16

[96] Lebuhn M, Weiss S, Munk B, Guebitz GM. Microbiology and molecular biology tools for biogas process analysis, diagnosis and control. 2015. p. 1-40

[97] De Vrieze J, Pinto AJ, Sloan WT, Ijaz UZ. The active microbial community more accurately reflects the anaerobic digestion process: 16S rRNA (gene) sequencing as a predictive tool. Microbiome. 2018;**6**:63. DOI: 10.1186/s40168-018-0449-9

[98] Hassa J, Maus I, Off S, Puehler A, Scherer P, Klocke M, et al. Metagenome, metatranscriptome, and metaproteome approaches unraveled compositions and functional relationships of microbial communities residing in biogas plants. Applied Microbiology and Biotechnology. 2018;**102**:5045-5063. DOI: 10.1007/s00253-018-8976-7

[99] Maus I, Rumming M, Bergmann I, Heeg K, Pohl M, Nettmann E, et al. Characterization of Bathyarchaeota genomes assembled from metagenomes of biofilms residing in mesophilic and thermophilic biogas reactors. Biotechnology for Biofuels. 2018;**11**:167. DOI: 10.1186/s13068-018-1162-4

[100] De Vrieze J, Gildemyn S, Vilchez-Vargas R, Jáuregui R, Pieper DH, Verstraete W, et al. Inoculum selection is crucial to ensure operational stability in anaerobic digestion. Applied Microbiology and Biotechnology. 2015;**99**:189-199. DOI: 10.1007/s00253-014-6046-3

[101] Raposo F, Fernandez-Cegri V, De la Rubia MA, Borja R, Beline F, Cavinato C, et al. Biochemical methane potential (BMP) of solid organic substrates: Evaluation of anaerobic biodegradability using data from an international interlaboratory study. Journal of Chemical Technology and Biotechnology. 2011;**86**:1088-1098. DOI: 10.1002/jctb.2622

[102] Han S, Liu YF, Zhang SC, Luo G. Reactor performances and microbial communities of biogas reactors: Effects of inoculum sources. Applied Microbiology and Biotechnology. 2016;**100**:987-995. DOI: 10.1007/s00253-015-7062-7

[103] Wojcieszak M, Pyzik A, Poszytek K, Krawczyk PS, Sobczak A, Lipinski L, et al. Adaptation of methanogenic inocula to anaerobic digestion of maize silage. Frontiers in Microbiology. 2017;**8**:1881. DOI: 10.3389/fmicb.2017.01881

[104] Satpathy P, Steinigeweg S, Cypionka H, Engelen B. Different substrates and starter inocula govern microbial community structures in biogas reactors. Environmental Technology. 2016;**37**:1441-1450. DOI: 10.1080/09593330.2015.1118559

[105] Parra-Orobio BA, Donoso-Bravo A, Ruiz-Sanchez JC, Valencia-Molina KJ, Torres-Lozada P. Effect of inoculum on the anaerobic digestion of food waste

accounting for the concentration of trace elements. Waste Management. 2018;**71**:342-349. DOI: 10.1016/j.wasman.2017.09.040

[106] De Vrieze J, Saunders AM, He Y, Fang J, Nielsen PH, Verstraete W, et al. Ammonia and temperature determine potential clustering in the anaerobic digestion microbiome. Water Research. 2015;**75**:312-323. DOI: 10.1016/j.watres.2015.02.025

[107] Sundberg C, Al-Soud WA, Larsson M, Alm EJ, Yekta SS, Svensson BH, et al. 454 pyrosequencing analyses of bacterial and archaeal richness in 21 full-scale biogas digesters. FEMS Microbiology Ecology. 2013;**85**:612-626. DOI: 10.1111/1574-6941.12148

[108] Koch K, Lippert T, Drewes JE. The role of inoculum's origin on the methane yield of different substrates in biochemical methane potential (BMP) tests. Bioresource Technology. 2017;**243**:457-463. DOI: 10.1016/j.biortech.2017.06.142

[109] Moreno-Andrade I, Buitron G. Influence of the origin of the inoculum on the anaerobic biodegradability test. Water Science and Technology. 2004;**49**:53-59. DOI: 10.2166/wst.2004.0017

[110] Rosenfeld JS. Functional redundancy in ecology and conservation. Oikos. 2002;**98**:156-162. DOI: 10.1034/j.1600-0706.2002.980116.x

[111] Lins P, Reitschuler C, Illmer P. *Methanosarcina* spp., the key to relieve the start-up of a thermophilic anaerobic digestion suffering from high acetic acid loads. Bioresource Technology. 2014;**152**:347-354. DOI: 10.1016/j.biortech.2013.11.020

[112] Liu T, Sun L, Nordberg A, Schnurer A. Substrate-induced response in biogas process performance and microbial community relates back to inoculum source. Microorganisms. 2018;**6**:E80. DOI: 10.3390/microorganisms6030080

[113] Hagen LH, Vivekanand V, Linjordet R, Pope PB, Eijsink VGH, Horn SJ. Microbial community structure and dynamics during co-digestion of whey permeate and cow manure in continuous stirred tank reactor systems. Bioresource Technology. 2014;**171**:350-359. DOI: 10.1016/j.biortech.2014.08.095

[114] Levén L, Eriksson A, Schnürer A. Effect of process temperature on bacterial and archaeal communities in two methanogenic bioreactors treating organic household waste. FEMS Microbiology Ecology. 2007;**59**:683-693. DOI: 10.1111/j.1574-6941.2006.00263.x

[115] Labatut RA, Angenent LT, Scott NR. Conventional mesophilic vs. thermophilic anaerobic digestion: A trade-off between performance and stability? Water Research. 2014;**53**:249-258. DOI: 10.1016/j.watres.2014.01.035

[116] Pap B, Györkei Á, Boboescu IZ, Nagy IK, Bíró T, Kondorosi É, et al. Temperature-dependent transformation of biogas-producing microbial communities points to the increased importance of hydrogentrophic methanogenesis under thermophilic operation. Bioresource Technology. 2015;**177**:375-380. DOI: 10.1016/j.biortech.2014.11.021

[117] Mei R, Nobu MK, Narihiro T, Kuroda K, Sierra JM, Wu Z, et al. Operation-driven heterogeneity and overlooked feed-associated populations in global anaerobic digester microbiome. Water Research. 2017;**124**:77-84. DOI: 10.1016/j.watres.2017.07.050

[118] Sun L, Pope PB, Eijsink VGH, Schnürer A. Characterization of microbial community structure during continuous anaerobic digestion of

straw and cow manure. Microbial Biotechnology. 2015;**8**:815-827. DOI: 10.1111/1751-7915.12298

[119] Moestedt J, Nordell E, Schnürer A. Comparison of operational strategies for increased biogas production from thin stillage. Journal of Biotechnology. 2014;**175**:22-30. DOI: 10.1016/j.jbiotec.2014.01.030

[120] Westerholm M, Müller B, Isaksson S, Schnürer A. Trace element and temperature effects on microbial communities and links to biogas digester performance at high ammonia levels. Biotechnology for Biofuels. 2015;**8**:1-19. DOI: 10.1186/s13068-015-0328-6

[121] Hupfauf S, Plattner P, Wagner AO, Kaufmann R, Insam H, Podmirseg SM. Temperature shapes the microbiota in anaerobic digestion and drives efficiency to a maximum at 45 °C. Bioresource Technology. 2018;**269**:309-318. DOI: 10.1016/j.biortech.2018.08.106

[122] Dhaked RK, Singh P, Singh L. Biomethanation under psychrophilic conditions. Waste Management. 2010;**30**:2490-2496. DOI: 10.1016/j.wasman.2010.07.015

[123] Lin Q, De Vrieze J, He G, Li X, Li J. Temperature regulates methane production through the function centralization of microbial community in anaerobic digestion. Bioresource Technology. 2016;**216**:150-158. DOI: 10.1016/j.biortech.2016.05.046

[124] Chen H, Chang S. Impact of temperatures on microbial community structures of sewage sludge biological hydrolysis. Bioresource Technology. 2017;**245**:502-510. DOI: 10.1016/j.biortech.2017.08.143

[125] Hwu C, Lettinga G. Acute toxicity of oleate to acetate-utilizing methanogens in mesophilic and thermophilic anaerobic sludges. Enzyme and Microbial Technology. 1997;**21**:297-301. DOI: 10.1016/S0141-0229(97)00050-1

[126] Levén L, Nyberg K, Schnürer A. Conversion of phenols during anaerobic digestion of organic solid waste—A review of important microorganisms and impact of temperature. Journal of Environmental Management. 2011;**95**:99-103. DOI: 10.1016/j.jenvman.2010.10.021

[127] Lindorfer H, Waltenberg R, Köller K, Braun R, Kirchmayr R. New data on temperature optimum and temperature changes in energy crop digesters. Bioresource Technology. 2008;**99**:7011-7019

[128] Moestedt J, Nordell E, Schnürer A. Comparison of operating strategies for increased biogas production from thin stillage. Journal of Biotechnology. 2014;**175**:22-30. DOI: 10.1016/j.jbiotec.2014.01.030

[129] Risberg K, Sun L, Levén L, Horn SJ, Schnürer A. Biogas production from wheat straw and manure—Impact of pretreatment and process operating parameters. Bioresource Technology. 2013;**149**:232-237

[130] Ziembinska-Buczynska A, Banach A, Bacza T, Pieczykolan M. Diversity and variability of methanogens during the shift from mesophilic to thermohilic conditions while biogas production. World Journal of Microbiology and Biotechnology. 2014;**30**:3047-3053

[131] Drosg B. Process Monitoring in Biogas Plants. IEA Bioenergy; 2013. Technical Brochure

[132] Kim MS, Kim DH, Yun YM. Effect of operation temperature on anaerobic digestion of food waste: Performance and microbial analysis. Fuel. 2017;**209**:598-605. DOI: 10.1016/j.fuel.2017.08.033

[133] Chapleur O, Mazeas L, Godon J, Bouchez T. Asymmetrical response of anaerobic digestion microbiota to temperature changes. Applied Microbiology and Biotechnology. 2016;**100**:1445-1457. DOI: 10.1007/s00253-015-7046-7

[134] Moset V, Poulsen M, Wahid R, Hojberg O, Moller HB. Mesophilic versus thermophilic anaerobic digestion of cattle manure: methane productivity and microbial ecology. Microbial Biotechnology. 2015;**8**:787-800

[135] Jang HM, Ha JH, Kim M, Kim J, Kim YM, Park JM. Effect of increased load of high-strength food wastewater in thermophilic and mesophilic anaerobic co-digestion of waste activated sludge on bacterial community structure. Water Research. 2016;**99**:140-148. DOI: 10.1016/j.watres.2016.04.051

[136] Theuerl S, Klang J, Heiermann M, De Vrieze J. Marker microbiome clusters are determined by operational parameters and specific key taxa combinations in anaerobic digestion. Bioresource Technology. 2018;**263**:128-135. DOI: 10.1016/j.biortech.2018.04.111

[137] Ahlberg-Eliasson K, Liu T, Nadeau E, Schnurer A. Forage types and origin of manure in codigestion affect methane yield and microbial community structure. Grass and Forage Science. 2018;**73**:740-757. DOI: 10.1111/gfs.12358

[138] Chen S, Cheng H, Wyckoff KN, He Q. Linkages of *Firmicutes* and *Bacteroidetes* populations to methanogenic process performance. Journal of Industrial Microbiology & Biotechnology. 2016;**43**:771-781. DOI: 10.1007/s10295-016-1760-8

[139] Güllert S, Fischer MA, Turaev D, Noebauer B, Ilmberger N, Wemheuer B, et al. Deep metagenome and metatranscriptome analyses of microbial communities affiliated with an industrial biogas fermenter, a cow rumen, and elephant feces reveal major differences in carbohydrate hydrolysis strategies. Biotechnology for Biofuels. 2016;**9**:121. DOI: 10.1186/s13068-016-0534-x

[140] Stolze Y, Bremges A, Rumming M, Henke C, Maus I, Pühler A, et al. Identification and genome reconstruction of aundant distinct taxa in microbiomes from one thermophilic and three mesophilic production-scale biogas plants. Biotechnology for Biofuels. 2016;**9**:156. DOI: 10.1186/s13068-016-0565-3

[141] Zamanzadeh M, Hagen LH, Svensson K, Linjordet R, Horn SJ. Anaerobic digestion of food waste—Effect of recirculation and temperature on performance and microbiology. Water Research. 2016;**96**:246-254. DOI: 10.1016/j.watres.2016.03.058

[142] Guo X, Wang C, Sun F, Shu W, Wu W. A comparison of microbial characteristics between the thermophilic and mesophilic anaerobic digesters exposed to elevated food waste loadings. Bioresource Technology. 2014;**152**:420-428. DOI: 10.1016/j.biortech.2013.11.012

[143] Tian Z, Zhang Y, Li YF, Chi Y, Yang M. Rapid establishment of thermophilic anaerobic microbial community during the one-step startup of thermophilic anaerobic digestion from a mesophilic digester. Water Research. 2015;**69**:9-19. DOI: 10.1016/j.watres.2014.11.001

[144] Maus I, Cibis KG, Bremges A, Stolze Y, Wibberg D, Blom J, et al. Genomic characterization of *Defluviitoga tunisiensis* L3, a key hydrolytic bacterium in a thermophilic biogas plant. Journal of Biotechnology. 2016;**232**:50-60. DOI: 10.1016/j.jbiotec.2016.05.001

[145] Balk M, Weijma J, Stams AJM. *Thermotoga lettingae* sp. nov., a novel thermophilic, methanol-degrading

bacterium isolated from a themophilic anaerobic reactor. International Journal of Systematic and Evolutionary Microbiology. 2002;**52**:1361-1368. DOI: 10.1099/00207713-52-4-1361

[146] Angelidaki I, Ellegaard L, Ahring BK. A mathematical model for dynamic simulation of anaerobic digestion of complex substrates: Focusing on ammonia inhibition. Biotechnology and Bioengineering. 1993;**42**:159-166

[147] Fontana A, Patrone V, Puglisi E, Morelli L, Bassi D, Garuti M, et al. Effects of geographic area, feedstock, temperature, and operating time on microbial communities of six full-scale biogas plants. Bioresource Technology. 2016;**218**:980-990. DOI: 10.1016/j.biortech.2016.07.058

[148] Yu D, Kurola JM, Lähde K, Kymäläinen M, Sinkkonen A, Romantschuk M. Biogas production and methanogenic archaeal community in mesophilic and thermophilic anaerobic co-digestion processes. Journal of Environmental Management. 2014;**143**:54-60. DOI: 10.1016/j.jenvman.2014.04.025

[149] Yin DM, Westerholm M, Qiao W, Bi SJ, Wandera SM, Fan R, et al. An explanation of the methanogenic pathway for methane production in anaerobic digestion of nitrogen-rich materials under mesophilic and thermophilic conditions. Bioresource Technology. 2018;**264**:42-50. DOI: 10.1016/j.biortech.2018.05.062

[150] Zhao J, Westerholm M, Qiao W, Yin D, Bi S, Jiang M, et al. Impact of temperature and substrate concentration on degradation rates of acetate, propionate and hydrogen and their links to microbial community structure. Bioresource Technology. 2018;**256**:44-52. DOI: 10.1016/j.biortech.2018.01.150

[151] Bousková A, Dohányos M, Schmidt JE, Angelidaki I. Strategies for changing temperature from mesophilic to thermophilic conditions in anaerobic CSTR reactors treating sewage sludge. Water Research. 2005;**39**:1481-1488. DOI: 10.1016/j.watres.2004.12.042

[152] Dinsdale RM, Hawkes FR, Hawkes DL. Comparision of mesophilic and thermophilic upflow anaerobic sludge blanket reactors treating instant coffee production wastewater. Water Research. 1997;**31**:163-169. DOI: 10.1016/S0043-1354(96)00233-3

[153] Patinvoh RJ, Osadolor OA, Chandolias K, Horváth S, Taherzadeh MJ. Innovative pretreatment strategies for biogas production. Bioresource Technology. 2017;**224**:13-24. DOI: 10.1016/j.biortech.2016.11.083

[154] Noike T, Endo G, Chang JE, Yaguchi JI, Matsumoto JI. Characteristics of carbohydrate degradation and the rate-limiting step in anaerobic digestion. Biotechnology and Bioengineering. 1985;**27**:1482-1489

[155] Kelessidis A, Stasinakis AS. Comparative study of the methods used for treatment and final disposal of sewage sludge in European countries. Waste Management. 2012;**32**:1186-1195. DOI: 10.1016/j.wasman.2012.01.012

[156] Rajesh Banu J, Kavitha S, Tamilarasan K. Modelling for anaerobic process. In: Purohit HJ, Kalia V, Vaidya A, Khardenavis A, editors. Optimization and Applicability of Bioprocesses. Singapore: Springer; 2017. pp. 283-304. DOI: 10.1007/978-981-10-6863-8_14

[157] Rajesh Banu J, Kavitha S. Various sludge pretreatments: Their impact on biogas generation. In: Singh L, Kalia VC, editors. Waste Biomass Management—A Holistic Approach. Cham: Springer; 2017. pp. 39-71. DOI: 10.1007/978-3-319-49595-8_3

[158] Kumari D, Singh R. Pretreatment of lignocellulosic wastes for biofuel

production: A critical review. Renewable and Sustainable Energy Reviews. 2018;**90**:877-891. DOI: 10.1016/j.rser.2018.03.111

[159] Park SK, Jang HM, Ha JH, Park JM. Sequential sludge digestion after diverse pre-treatment conditions: sludge removal, methane production and microbial community changes. Bioresource Technology. 2014;**162**:331-340. DOI: 10.1016/j.biortech.2014.03.152

[160] Gagliano MC, Braguglia CM, Gianico A, Mininni G, Nakamura K, Rossetti S. Thermophilic anaerobic digestion of thermal pretreated sludge: Role of microbial community structure and correlation with process performances. Water Research. 2015;**68**:498-509. DOI: 10.1016/j.watres.2014.10.031

[161] Wong MT, Zhang D, Li J, Hui RKH, Tun HM, Brar MS, et al. Towards a metagenomic understanding on enhanced biomethane production from waste activated sludge after pH 10 pretreatment. Biotechnology for Biofuels. 2013;**6**:38. DOI: 10.1186/1754-6834-6-38

[162] Westerholm M, Crauwels S, Houtmeyers S, Meerbergen K, Van Geel M, Lievens B, et al. Microbial community dynamics linked to enhanced substrate availability and biogas production of electrokinetically pre-treated waste activated sludge. Bioresource Technology. 2016;**218**:761-770. DOI: 10.1016/j.biortech.2016.07.029

[163] Westerholm M, Crauwels S, Van Geel M, Dewil R, Lievens B, Appels L. Microwave and ultrasound pre-treatments influence microbial community structure and digester performance in anaerobic digestion of waste activated sludge. Applied Microbiology and Biotechnology. 2016;**100**:5339-5352. DOI: 10.1007/s00253-016-7321-2

[164] Sanz JL, Rojas P, Morato A, Mendez L, Ballesteros M, Gonzáles-Fernández C. Microbial communities of biomethanization digesters fed with raw and heat pre-treated microalgae biomasses. Chem. 2017;**168**:1013-1021. DOI: 10.1016/j.chemosphere.2016.10.109

[165] Xia Y, Yang C, Zhang T. Microbial effects of part-stream low-frequency ultrasonic pretreatment on sludge anaerobic digestion as revealed by high-throughput sequencing-based metagenomics and metatranscriptomics. Biotechnology for Biofuels. 2018;**11**:47. DOI: 10.1186/s13068-018-1042-y

[166] Wang TT, Sun ZY, Huang YL, Tan L, Tang YQ, Kida K. Biogas production from distilled grain waste by thermophilic dry anaerobic digestion: Pretreatment of feedstock and dynamics of microbial community. Applied Biochemistry and Biotechnology. 2018;**184**:685-702. DOI: 10.1007/s12010-017-2557-6

[167] Monlau F, Sambusiti C, Barakat A, Quemeneur M, Trably E, Steyer JP, et al. Do furanic and phenolic compounds of lignocellulosic and algae biomass hydrolyzate inhibit anaerobic mixed cultures? A comprehensive review. Biotechnology Advances. 2014;**32**:934-951. DOI: 10.1016/j.biotechadv.2014.04007

[168] Poirier S, Bize A, Bureau C, Bouchez T, Chapleur O. Community shifts within anaerobic digestion microbiota facing phenol inhibition: Towards early warning microbial indicators? Water Research. 2016;**100**:296-305. DOI: 10.1016/j.watres.2016.05.041

[169] Na JG, Lee MK, Yun YM, Moon C, Kim MS, Kim DH. Microbial community analysis of anaerobic

granules in phenol-degrading UASB by next generation sequencing. Biochemical Engineering Journal. 2016;**112**:241-248. DOI: 10.1016/j.bej.2016.04.030

[170] Ozgun H, Dereli RK, Ersahin ME, Kinaci C, Spanjers H, van Lier JB. A review of anaerobic membrane bioreactors for municipal wastewater treatment: Integration options, limitations and expectations. Separation and Purification Technology. 2013;**118**:89-104. DOI: 10.1016/j.seppur.2013.06.036

[171] Karadag D, Koroglu OE, Ozkaya B, Cakmakci M. A review on anaerobic biofilm reactors for the treatment of dairy industry wastewater. Process Biochemistry. 2015;**50**:262-271. DOI: 10.1016/j.procbio.2014.11.005

[172] Xu R, Yang Z-H, Zheng Y, Liu J-B, Xiong W-P, Zhang Y-R, et al. Organic loading rate and hydraulic retention time shape distinct ecological networks of anaerobic digestion related microbiome. Bioresource Technology. 2018;**262**:184-193. DOI: 10.1016/j.biortech.2018.04.083

[173] Bonk F, Popp D, Weinrich S, Straeuber H, Kleinsteuber S, Harms H, et al. Intermittent fasting for microbes: How discontinuous feeding increases functional stability in anaerobic digestion. Biotechnology for Biofuels. 2018;**11**:274. DOI: 10.1186/s13068-018-1279-5

[174] Liu C, Li H, Zhang YY, Si DD, Chen QW. Evolution of microbial community along with increasing solid concentration during high-solids anaerobic digestion of sewage sludge. Bioresource Technology. 2016;**216**:87-94. DOI: 10.1016/j.biortech.2016.05.048

[175] Liang X, Whitham JM, Holwerda EK, Shao X, Tian L, Wu Y-W, et al. Development and characterization of stable anaerobic thermophilic methanogenic microbiomes fermenting switchgrass at decreasing residence times. Biotechnology for Biofuels. 2018;**11**:243. DOI: 10.1186/s13068-018-1238-1

[176] Dennehy C, Lawlor PG, Gardiner GE, Jiang Y, Cormican P, McCabe MS, et al. Process stability and microbial community composition in pig manure and food waste anaerobic co-digesters operated at low HRTs. Frontiers of Environmental Science & Engineering. 2017;**11**. DOI: 10.1007/s11783-017-0923-9

[177] Tonanzi B, Gallipoli A, Gianico A, Montecchio D, Pagliaccia P, Di Carlo M, et al. Long-term anaerobic digestion of food waste at semi-pilot scale: Relationship between microbial community structure and process performances. Biomass and Bioenergy. 2018;**118**:55-64. DOI: 10.1016/j.biombioe.2018.08.001

[178] Martins AD, Ferreira BO, Ribeiro NC, Martins R, Leite LR, Oliveira G, et al. Metagenomic analysis and performance of a mesophilic anaerobic reactor treating food waste at various load rates. Environmental Technology. 2017;**38**:2153-2163. DOI: 10.1080/09593330.2016.1247197

[179] Moestedt J, Müller B, Westerholm M, Schnürer A. Ammonia threshold for inhibition of anaerobic digestion of thin stillage and the importance of organic loading rate. Microbial Biotechnology. 2016;**9**:180-194. DOI: 10.1111/1751-7915.12330

[180] Leite AF, Janke L, Lv ZP, Harms H, Richnow HH, Nikolausz M. Improved monitoring of semi-continuous anaerobic digestion of sugarcane waste: Effects of increasing organic loading rate on methanogenic community dynamics. International Journal of Molecular

Sciences. 2015;**16**:23210-23226. DOI: 10.3390/ijms161023210

[181] Hori T, Haruta S, Sasaki D, Hanajima D, Ueno Y, Ogata A, et al. Reorganization of the bacterial and archaeal populations associated with organic loading conditions in a thermophilic anaerobic digester. Journal of Bioscience and Bioengineering. 2015;**119**:337-344. DOI: 10.1016/j.jbiosc.2014.09.003

[182] Lerm S, Kleyboecker A, Miethling-Graff R, Alawi M, Kasina M, Liebrich M, et al. Archaeal community composition affects the function of anaerobic co-digesters in response to organic overload. Waste Management. 2012;**32**:389-399. DOI: 10.1016/j.wasman.2011.11.013

[183] Ziels RM, Svensson BH, Sundberg C, Larsson M, Karlsson A, Yekta SS. Microbial rRNA gene expression and co-occurrence profiles associate with biokinetics and elemental composition in full-scale anaerobic digesters. Microbial Biotechnology. 2018;**11**:694-709. DOI: 10.1111/1751-7915.13264

[184] Xu R, Zhang K, Liu P, Khan A, Xiong J, Tian F, et al. A critical review on the interaction of substrate nutrient balance and microbial community structure and function in anaerobic co-digestion. Bioresource Technology. 2018;**247**:1119-1127. DOI: 10.1016/j.biortech.2017.09.095

[185] Sprott GD, Shaw KM, Jarell KF. Methanogenesis and the K^+ transport system are activated by divalent cations in ammonia-treated cells of *Methanospirillum hungatei*. The Journal of Biological Chemistry. 1985;**260**:9244-9250

[186] Sprott GD, Shaw KM, Jarell KF. Ammonia/potassium exchange in methanogenic bacteria. The Journal of Biological Chemistry. 1984;**259**:12602-12608

[187] Schnürer A, Nordberg A. Ammonia, a selective agent for methane production by syntrophic acetate oxidation at mesophilic temperature. Water Science and Technology. 2008;**57**:735-740. DOI: 10.2166/wst.2008.097

[188] Westerholm M, Dolfing J, Sherry A, Gray ND, Head IM, Schnürer A. Quantification of syntrophic acetate-oxidizing microbial communities in biogas processes. Environmental Microbiology Reports. 2011;**3**:500-505. DOI: 10.1111/j.1758-2229.2011.00249.x

[189] Westerholm M, Levén L, Schnürer A. Bioaugmentation of syntrophic acetate-oxidising culture in biogas reactors exposed to increasing levels of ammonia. Applied and Environmental Microbiology. 2012;**78**:7619-7625. DOI: 10.1128/AEM.01637-12

[190] Sun L, Müller B, Westerholm M, Schnürer A. Syntrophic acetate oxidation in industrial CSTR biogas digesters. Journal of Biotechnology. 2014;**171**:39-44. DOI: 10.1016/j.jbiotec.2013.11.016

[191] Lü F, Hao L, Guan D, Qi Y, Shao L, He P. Synergetic stress of acids and ammonium on the shift in the methanogenic pathways during thermophilic anaerobic digestion of organics. Water Research. 2013;**47**:2297-2306

[192] Fotidis IA, Karakashev D, Angelidaki I. The dominant acetate degradation pathway/methanogenic composition in full-scale anaerobic digesters operating under different ammonia levels. International Journal of Environmental Science and Technology. 2014;**11**: 2087-2094. DOI: 10.1007/s13762-013-0407-9

[193] Werner JJ, Garcia ML, Perkins SD, Yarasheski KE, Smith SR, Muegge B, et al. Microbial community dynamics and stability during an

ammonia-induced shift to syntrophic acetate oxidation. Applied and Environmental Microbiology. 2014;**80**:3375-3383. DOI: 10.1128/AEM.00166-14

[194] Hattori S, Kamagata Y, Hanada S, Shoun H. *Thermacetogenium phaeum* gen. nov., sp. nov., a strictly anaerobic, thermophilic, syntrophic acetate-oxidizing bacterium. International Journal of Systematic and Evolutionary Microbiology. 2000;**50**:1601-1609. DOI: 10.1099/00207713-50-4-1601

[195] Bhandari V, Gupta RS. Molecular signatures for the phylum (class) Thermotogae and a proposal for its division into three orders (*Thermotogales*, *Kosmotogales* ord. nov. and *Petrotogales* ord. nov.) containing four families (*Thermotogaceae*, *Fervidobacteriaceae* fam. nov., *Kosmotogaceae* fam. nov. and *Petrotogaceae* fam. nov.) and a new genus *Pseudothermotoga* gen. nov. with five new combinations. Antonie Van Leeuwenhoek. 2014;**105**:143-168. DOI: 10.1007/s10482-013-0062-7

[196] Westerholm M, Roos S, Schnürer A. *Tepidanaerobacter acetatoxydans* sp. nov., an anaerobic, syntrophic acetate-oxidizing bacterium isolated from two ammonium-enriched mesophilic methanogenic processes. Systematic and Applied Microbiology. 2011;**34**:260-266. DOI: 10.1016/j.syapm.2010.11.018

[197] Schnürer A, Schink B, Svensson BH. *Clostridium ultunense* sp. nov., a mesophilic bacterium oxidizing acetate in syntrophic association with a hydrogenotrophic methanogenic bacterium. International Journal of Systematic Bacteriology. 1996;**46**:1145-1152

[198] Westerholm M, Roos S, Schnürer A. *Syntrophaceticus schinkii* gen. nov., sp. nov., an anaerobic, syntrophic acetate-oxidizing bacterium isolated from a mesophilic anaerobic filter. FEMS Microbiology Letters. 2010;**309**:100-104. DOI: 10.1111/j.1574-6968.2010.02023.x

[199] Karakashev D, Batstone DJ, Trably E, Angelidaki I. Acetate oxidation is the dominant methanogenic pathway from acetate in the absence of *Methanosaetaceae*. Applied and Environmental Microbiology. 2006;**72**:5138-5141. DOI: 10.1128/AEM.00489-06

[200] Karlsson A, Einarsson P, Schnürer A, Eljertsson J, Svensson BH. Impact of trace element addition on degradation efficiency of volatile fatty acids, oleic acid and phenyl acetate and on microbial populations in a biogas digester. Journal of Bioscience and Bioengineering. 2012;**114**:446-452. DOI: 10.1016/j.jbiosc.2012.05.010

[201] Francisci DD, Kougias PG, Treu L, Campanaro S, Angelidaki I. Microbial diversity and dynamicity of biogas reactors due to radical changes of feedstock composition. Bioresource Technology. 2015;**176**:56-64. DOI: 10.1016/j.biortech.2014.10.126

[202] Bozan M, Akyol C, Ince O, Aydin S, Ince B. Application of next-generation sequencing methods for microbial monitoring of anaerobic digestion of lignocellulosic biomass. Applied Microbiology and Biotechnology. 2017;**101**:6849-6864. DOI: 10.1007/s00253-017-8438-7

[203] Mu H, Zhao C, Zhao Y, Li Y, Hua D, Zhang X, et al. Enhanced methane production by semi-continuous mesophilic co-digestion of potato waste and cabbage waste: Performance and microbial characteristics analysis. Bioresource Technology. 2017;**236**:68-76. DOI: 10.1016/j.biortech.2017.03.138

[204] Zealand AM, Mei R, Papachristodoulou P, Roskilly AP, Liu WT, Graham DW. Microbial community composition and diversity in rice straw

digestion bioreactors with and without dairy manure. Applied Microbiology and Biotechnology. 2018;**102**:8599-8612. DOI: 10.1007/s00253-018-9243-7

[205] Wang M, Zhou J, Yuan Y-X, Dai Y-M, Li D, Li Z-D, et al. Methane production characteristics and microbial community dynamics of mono-digestion and co-digestion using corn stalk and pig manure. International Journal of Hydrogen Energy. 2017;**42**:4893-4901. DOI: 10.1016/j.ijhydene.2016.10.144

[206] Li X, Liu YH, Zhang X, Ge CM, Piao RZ, Wang WD, et al. Evaluation of biogas production performance and dynamics of the microbial community in different straws. Journal of Microbiology and Biotechnology. 2017;**27**:524-534. DOI: 10.4014/jmb.1608.08062

[207] Ozbayram EG, Kleinsteuber S, Nikolausz M, Ince B, Ince O. Enrichment of lignocellulose-degrading microbial communities from natural and engineered methanogenic environments. Applied Microbiology and Biotechnology. 2018;**102**:1035-1043. DOI: 10.1007/s00253-017-8632-7

[208] Yu Q, Tian Z, Liu J, Zhou J, Yan Z, Yong X, et al. Biogas production and microbial community dynamics during the anaerobic digestion of rice straw at 39-50°C: A pilot study. Energy & Fuels. 2018;**32**:5157-5163. DOI: 10.1021/acs.energyfuels.7b04042

[209] Zhou J, Yang J, Yu Q, Yong X, Xie X, Zhang L, et al. Different organic loading rates on the biogas production during the anaerobic digestion of rice straw: A pilot study. Bioresource Technology. 2017;**244**:865-871. DOI: 10.1016/j.biortech.2017.07.146

[210] Ros M, JDS OF, Perez Murcia MD, Angeles Bustamante M, Moral R, Dolores Coll M, et al. Mesophilic anaerobic digestion of pig slurry and fruit and vegetable waste: Dissection of the microbial community structure. Journal of Cleaner Production. 2017;**156**:757-765. DOI: 10.1016/j.jclepro.2017.04.110

[211] Lu X, Rao S, Shen Z, Lee PKH. Substrate induced emergence of differenct active bacterial and archaeal assemblages during biomethane production. Bioresource Technology. 2013;**148**:517-524. DOI: 10.1016/j.biortech.2013.09.017

[212] Grosser A, Neczaj E. Enhancement of biogas production from sewage sludge by addition of grease trap sludge. Energy Conversion and Management. 2016;**125**:301-308. DOI: 10.1016/j.enconman.2016.05.089

[213] Martinez EJ, Gil MV, Fernandez C, Rosas JG, Gomez X. Anaerobic codigestion of sludge: Addition of butcher's fat waste as a cosubstrate for increasing biogas production. PLoS One. 2016;**11**:e0153139. DOI: 10.1371/journal.pone.0153139

[214] Zhu Z, Hsueh MK, He Q. Enhancing biomethanation of municipal waste sludge with grease trap waste as a co-substrate. Renewable Energy. 2011;**36**:1802-1807. DOI: 10.1016/j.renene.2010.11.014

[215] Ahlberg-Eliasson K, Nadeau E, Leven L, Schnurer A. Production efficiency of Swedish farm-scale biogas plants. Biomass and Bioenergy. 2017;**97**:27-37. DOI: 10.1016/j.biombioe.2016.12.002

[216] Amha YM, Sinha P, Lagman J, Gregori M, Smith AL. Elucidating microbial community adaptation to anaerobic co-digestion of fats, oils, and grease and food waste. Water Research. 2017;**123**:277-289. DOI: 10.1016/j.watres.2017.06.065

[217] Ziels RM, Beck DAC, Marti M, Gough HL, Stensel HD, Svensson BH. Monitoring the dynamics of syntrophic beta-oxidizing bacteria during anaerobic degradation

of oleic acid by quantitative PCR. FEMS Microbiology Ecology. 2015;**91**:fiv028. DOI: 10.1093/femsec/fiv028

[218] Ziels RM, Beck DAC, Stensel HD. Long-chain fatty acid feeding frequency in anaerobic codigestionimpacts syntrophic community structure and biokinetics. Water Research. 2017;**11**:218-229. DOI: 10.1016/j.watres.2017.03.060

[219] Ning Z, Zhang H, Li W, Zhang R, Liu G, Chen C. Anaerobic digestion of lipid-rich swine slaughterhouse waste: Methane production performance, long-chain fatty acids profile and predominant microorganisms. Bioresource Technology. 2018;**269**:426-433. DOI: 10.1016/j.biortech.2018.08.001

[220] Kougias PG, Treu L, Campanaro S, Zhu X, Angelidaki I. Dynamic functional characterization and phylogenetic changes due to Long Chain Fatty Acids pulses in biogas reactors. Science Reporter. 2016;**6**:28810. DOI: 10.1038/srep28810

[221] Sousa DZ, Pereira A, Stams AJM, Alves MM, Smidt H. Microbial communities involved in anaerobic degradation of unsaturated or saturated long-chain fatty acids. Applied and Environmental Microbiology. 2007;**73**:1054-1064. DOI: 10.1128/AEM.01723-06

[222] Baserba MG, Angelidaki I, Karakashev D. Effect of continuous oleate addition on microbial communities involved in anaerobic digestion process. Bioresource Technology. 2012;**106**:74-81. DOI: 10.1016/j.biortech.2011.12.020

[223] Shigematsu T, Tang Y, Mizuno Y, Kawaguchi H, Morimura S, Kida K. Microbial diversity of mesophilic methanogenic consortium that can degrade long-chain fatty acids in chemostat cultivation. Journal of Bioscience and Bioengineering. 2006;**102**:535-544. DOI: 10.1263/jbb.102.535

[224] Yekta SS, Ziels RM, Bjorn A, Skyllberg U, Ejlertsson J, Karlsson A, et al. Importance of sulfide interaction with iron as regulator of the microbial community in biogas reactors and its effect on methanogenesis, volatile fatty acids turnover, and syntrophic long-chain fatty acids degradation. Journal of Bioscience and Bioengineering. 2017;**123**:597-605. DOI: 10.1016/j.jbiosc.2016.12.003

[225] Regueiro L, Carballa M, Lema JM. Microbiome response to controlled shifts in ammonium and LCFA levels in co-digestion systems. Journal of Biotechnology. 2016;**220**:35-44. DOI: 10.1016/j.jbiotec.2016.01.006

[226] Razaviarani V, Buchanan ID. Anaerobic co-digestion of biodiesel waste glycerin with municipal wastewater sludge: Microbial community structure dynamics and reactor performance. Bioresource Technology. 2015;**182**:8-17. DOI: 10.1016/j.biortech.2015.01.095

[227] Jensen PD, Astals S, Lu Y, Devadas M, Batstone DJ. Anaerobic codigestion of sewage sludge and glycerol, focusing on process kinetics, microbial dynamics and sludge dewaterability. Water Research. 2014;**67**:355-366. DOI: 10.1016/j.watres.2014.09.024

[228] Nga Thi D, Hatta K, Kwon SH, Rollon AP, Nakasaki K. Changes in the microbial community during the acclimation stages of the methane fermentation for the treatment of glycerol. Biomass and Bioenergy. 2014;**68**:240-249. DOI: 10.1016/j.biombioe.2014.06.022

[229] Treu L, Campanaro S, Kougias PG, Zhu X, Angelidaki I. Untangling the effect of fatty acid addition at species level revealed different transcriptional responses of the biogas microbial community

members. Environmental Science & Technology. 2016;**50**:6079-6090. DOI: 10.1021/acs.est.6b00296

[230] Sousa DZ, Salvador AF, Ramos J, Guedes AP, Barbosa S, Stams AJM, et al. Activity and viability of methanogens in anaerobic digestion of unsaturated and saturated long-chain fatty acids. Applied and Environmental Microbiology. 2013;**79**:4239-4245. DOI: 10.1128/aem.00035-13

[231] Zhang C, Su H, Baeyens J, Tan T. Reviewing the anaerobic digestion of food waste for biogas production. Renewable and Sustainable Energy Reviews. 2014;**38**:383-392. DOI: 10.1016/j.rser.2014.05.038

[232] Moestedt J, Nordell E, Yekta SS, Lundgren J, Martí M, Sundberg C, et al. Effects of trace element addition on process stability during anaerobic co-digestion of OFMSW and slaughterhouse waste. Waste Management. 2015;**47**:11-20. DOI: 10.1016/j.wasman.2015.03.007

[233] Bayr S, Pakarinene O, Korppoo A, Liuksia S, Väisänen A, Kaparaju P, et al. Effect of additives on process stability of mesophilic anaerobic monodigestion of pig slaughterhouse waste. Bioresource Technology. 2012;**120**:106-113. DOI: 10.1016/j.biortech.2012.06.009

[234] Eftaxias A, Diamantis V, Aivasidis A. Anaerobic digestion of thermal pre-treated emulsified slaughterhouse wastes (TESW): Effect of trace element limitation on process efficiency and sludge metabolic properties. Waste Management. 2018;**76**:357-363. DOI: 10.1016/j.wasman.2018.02.032

[235] Lebuhn M, Liu F, Heuwinkel H, Gronauer A. Biogas production from mono-digestion of maize silage-long-term process stability and requirements. Water Science and Technology. 2008;**58**:1645-1651. DOI: 10.2166/wst.2008.495

[236] Gustavsson J, Svensson BH, Karlsson A. The feasibility of trace element supplementation for stable operation of wheat stillage-fed biogas tank reactors. Water Science and Technology. 2011;**64**:320-325. DOI: 0273-1223

[237] Nordell E, Nilsson B, Nilsson Påledal S, Karisalmi K, Moestedt J. Co-digestion of manure and industrial waste—The effects of trace element addition. Waste Management. 2016;**47**:21-27. DOI: 10.1016/j.wasman.2015.02.032

[238] Molaey R, Bayrakdar A, Sürmeli RO, Calli B. Influence of trace element supplementation on anaerobic digestion ofchicken manure: Linking process stability to methanogenic population dynamics. Journal of Cleaner Production. 2018;**181**:794-800

[239] Yekta SS, Svensson BH, Björn A, Skyllberg U. Thermodynamic modeling of iron and trace metal solubility and speciation under sulfidic and ferruginous conditions in full scale continuou stirred tank biogas reactors. Applied Geochemistry. 2014;**47**:61-73. DOI: 10.1016/j.apgeochem.2014.05.001

[240] Gustavsson J, Shakeri Yekta S, Sundberg C, Karlsson A, Eljertsson J, Skyllberg U, et al. Bioavailability of cobalt and nickel during anaerobic digestion of sulfur-rich stillage for biogas formation. Applied Energy. 2013;**112**:473-477. DOI: 10.1016/j.apenergy.2013.02.009

[241] Pham Minh T, Ketheesan B, Yan Z, Stuckey DC. Trace metal speciation and bioavailability in anaerobic digestion: A review. Biotechnology Advances. 2016;**34**:122-136. DOI: 10.1016/j.biotechadv.2015.12.006

[242] Speece RE. A survey of municipal anaerobic sludge digesters and diagnostic activity assays. Water Research. 1988;**22**:365-372. DOI: 10.1016/S0043-1354(88)90260-6

[243] de Diego-Diaz B, Alejandro Cerdan JM, Javier Penas F, Fernandez-Rodriguez J. Impact of supplementary nutrients on codigestion of agricultural waste: Study of temperatures. Food and Bioproducts Processing. 2018;**110**: 120-125. DOI: 10.1016/j.fbp.2018.05.003

[244] Hartmann T, Schwanhold N, Leimkuehler S. Assembly and catalysis of molybdenum or tungsten-containing formate dehydrogenases from bacteria. Biochimica et Biophysica Acta. 2015;**1854**:1090-1100. DOI: 10.1016/j.bbapap.2014.12.006

[245] Demirel B, Scherer P. Trace element requirements of agricultural biogas digesters during biological conversion of renewable biomass to methane. Biomass and Bioenergy. 2011;**35**:992-998. DOI: 10.1016/j.biombioe.2010.12.022

[246] Azim AA, Pruckner C, Kolar P, Taubner R-S, Fino D, Saracco G, et al. The physiology of trace elements in biological methane production. Bioresource Technology. 2017;**241**: 775-786. DOI: 10.1016/j.biortech.2017.05.211

[247] Molaey R, Bayrakdar A, Sürmeli RO, Calli B. Influence of trace element supplementation on anaerobic digestion of chicken manure: Linking process stability to methanogenic population dynamics. Journal of Cleaner Production. 2018;**181**:794-800. DOI: 10.1016/j.jclepro.2018.01.264

[248] Wintsche B, Jehmlich N, Popp D, Harms H, Kleinstauber S. Metabolic adaptation of methanogens in anaerobic digesters upon trace element limitation. Frontiers in Microbiology. 2018;**9**:405. DOI: 10.3389/fmicb.2018.00405

[249] Neubeck A, Sjöberg S, Price A, Callac N, Schnürer A. Effect of nickel levels on hydrogen partial pressure and methane production in methanogens. PLoS One. 2016;**11**:e0168357. DOI: 10.1371/journal.pone.0168357

[250] Nzila A. Mini review: Update on bioaugmentation in anaerobic processes for biogas production. Anaerobe. 2017;**46**:3-17. DOI: 10.1016/j.anaerobe.2016.11.007

[251] Öner BE, Akyol C, Bozan M, Ince O, Aydin S, Ince B. Bioaugmentation with *Clostridium thermocellum* to enhance the anaerobic biodegradation of lignocellulosic agricultural residues. Bioresource Technology. 2018;**249**: 620-625. DOI: 10.1016/j.biortech.2017.10.040

[252] Speda J, Johansson MA, Odnell A, Karlsson M. Enhanced biomethane production rate and yield from lignocellulosic ensiled forage ley by in situ anaerobic digestion treatment with endogenous cellulolytic enzymes. Biotechnology for Biofuels. 2017;**10**: 129. DOI: 0.1186/s13068-017-0814-0

[253] Peng X, Börner RA, Nges IA, Liu J. Impact of bioaugmentation on biochemical methane potential for wheat straw with addition of *Clostridium cellulolyticum*. Bioresource Technology. 2014;**152**:567-571. DOI: 10.1016/j.biortech.2013.11.067

[254] Zhang J, Guo R, Qiu Y, Qiao J, Yuan X, Shi X, et al. Bioaugmentation with an acetate-type fermentation bacterium *Acetobacteroides hydrogenigenes* improves methane production from corn straw. Bioresource Technology. 2015;**179**:306-313. DOI: 10.1016/j.biortech.2014.12.022

[255] Poszytek K, Ciezkowska M, Sklodowska A, Drewniak L. Microbial Consortium with High Cellulolytic Activity (MCHCA) for enhanced biogas production. Frontiers in Microbiology. 2016;**7**:324. DOI: 10.3389/fmicb.2016.00324

[256] Odnell A, Recktenwald M, Stensén K, Jonsson B, Karlsson M. Activity, life time and effect of hydrolytic enzymes for

enhanced biogas production from sludge anaerobic digestion. Water Research. 2016;**103**:462-471. DOI: 10.1016/j.watres.2016.07.064

[257] Azman S, Khadem AF, Plugge CM, Stams AJM, Bec S, Zeeman G. Effect of humic acid on anaerobic digestion of cellulose and xylan in completely stirred tank reactors: inhibitory effect, mitigation of the inhibition and the dynamics of the microbial communities. Applied Microbiology and Biotechnology. 2017;**101**:889-901. DOI: 10.1007/s00253-016-8010-x

[258] Cirne DG, Paloumet X, Bjornsson L, Alves MM, Mattiasson B. Anaerobic digestion of lipid-rich waste—Effects of lipid concentration. Renewable Energy. 2007;**32**:965-975. DOI: 10.1016/j.renene.2006.04.003

[259] Silva SA, Cavaleiro AJ, Pereira MA, Stams AJM, Alves MM, Sousa DZ. Long-term acclimation of anaerobic sludges for high-rate methanogenesis from LCFA. Biomass and Bioenergy. 2014;**67**:297-303. DOI: 10.1016/j.biombioe.2014.05.012

[260] Sasaki K, Morita M, Sasaki D, Nagaoka J, Matsumoto N, POhmura N, et al. Syntrophic degradation of proteinaceous materials by the thermophilic strains *Coprothermobacter proteolyticus* and *Methanothermobacter thermautotrophicus*. Journal of Bioscience and Bioengineering. 2011;**112**:469-472. DOI: 10.1016/j.jbiosc.2011.07.003

[261] Fotidis IA, Karakashev D, Angelidaki I. Bioaugmentation with an acetate-oxidising consortium as a tool to tackle ammonia inhibition of anaerobic digestion. Bioresource Technology. 2013;**146**:57-62. DOI: 10.1016/j.biortech.2013.07.041

[262] Fotidis IA, Wang H, Fiedel N-R, Luo G, Karakashev DB. Bioaugmentation as a solution to increase methane production from an ammonia-rich substrate. Environmental Science & Technology. 2014;**48**:7669-7676. DOI: 10.1021/es5017075

Chapter 3

Searching for Metabolic Pathways of Anaerobic Digestion: A Useful List of the Key Enzymes

Anna Sikora, Anna Detman, Damian Mielecki, Aleksandra Chojnacka and Mieczysław Błaszczyk

Abstract

The general scheme of anaerobic digestion is well known. It is a complex process promoted by the interaction of many groups of microorganisms and has four major steps: hydrolysis, acidogenesis, acetogenesis, and methanogenesis. The aim of the study was to prepare a systematized list of the selected enzymes responsible for the key pathways of anaerobic digestion based on the Kyoto Encyclopedia of Genes and Genomes database resource. The list contains (i) key groups of hydrolases involved in the process of degradation of organic matter; (ii) the enzymes catalyzing reactions leading to pyruvate formation; (iii) the enzymes of metabolic pathways of further pyruvate transformations; (iv) the enzymes of glycerol transformations; (v) the enzymes involved in transformation of gaseous or nongaseous products of acidic fermentations resulting from nonsyntrophic nutritional interactions between microbes; (vi) the enzymes of amino acid fermentations; (vii) the enzymes involved in acetogenesis; and (viii) the enzymes of the recognized pathways of methanogenesis. Searching for the presence and activity of the enzymes as well as linking structure and function of microbial communities allows to develop a fundamental understanding of the processes, leading to methane production. In this contribution, the present study is believed to be a piece to the enzymatic road map of anaerobic digestion research.

Keywords: anaerobic digestion, enzymes, hydrolysis, acidogenesis, acetogenesis, methanogenesis, syntrophy, metabolic pathways

1. Introduction

Anaerobic digestion (AD), whose final products are methane and carbon dioxide, is a common process in natural anoxic environments such as water sediments, wetlands, or marshlands. The environments have to be rich in organic matter and poor with other electron acceptors such as nitrate, compounds containing oxidized forms of metals, and sulfate. AD is also common in landfills and wastewater treatment plants and was used by man to produce biogas from waste biomass as an alternative energy source.

AD is a complex process that requires the metabolic interaction of many groups of microorganisms responsible for four closely related major steps. The first one is hydrolysis of complex organic polymers (e.g., polysaccharides, lipids, proteins) to monomers (sugars, fatty acids, amino acids). The second step is acidogenesis that results in formation of hydrogen and carbon dioxide as well as nongaseous fermentation products, that is, low-molecular-weight organic acids and alcohols. These products are further oxidized to hydrogen, carbon dioxide, and acetate in acetogenic step that involves mainly syntrophic degradation of nongaseous fermentation products. The fourth step is methanogenesis. Three groups of substrates for methane production and three types of methanogenic pathways are known: splitting of acetate (aceticlastic/acetotrophic methanogenesis); reduction of CO_2 with H_2 or formate and rarely ethanol or secondary alcohols as electron donors (hydrogenotrophic methanogenesis); and reduction of methyl groups of methylated compounds such as methanol, methylated amines, or methylated sulfides (hydrogen-dependent and hydrogen-independent methylotrophic methanogenesis). The two last steps, acetogenesis and methanogenesis, are closely related and involve syntrophic associations between hydrogen-producing acetogenic bacteria and hydrogenotrophic methanogens (**Figure 1**) [1–5].

Recently, there has been a rapid development in culture-independent techniques (meta-omics approaches such as metagenomics, metatranscriptomics, metaproteomics, metabolomics) for exploring microbial communities, which have led to a new insight into their structure and function in both natural environments and anaerobic digesters. The current trends involve the combined use of meta-omic approaches and detailed reactor performance data as well as isotope labeling techniques that allow us to develop a fundamental understanding of the processes occurring in AD. Those activities are aimed to improve

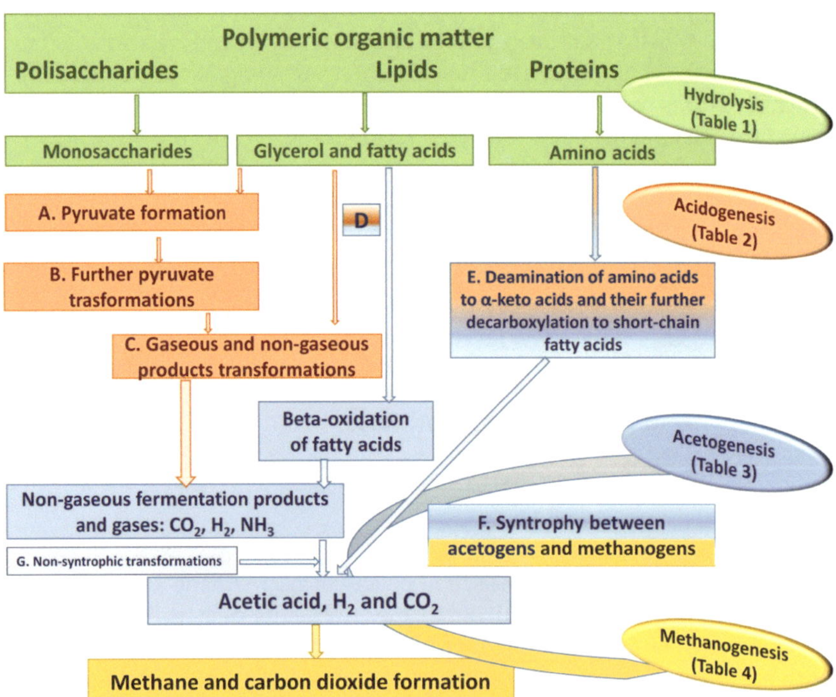

Figure 1.
A scheme of anaerobic digestion of organic matter. Enzymes catalysing specific reactions of AD are presented in ***Tables 1–4****. Thus in Figure 1 there are the links to* ***Tables 1–4****. Furthermore, background colours in the Figure correspond to the background colours of the title rows in the* ***Tables 1–4****: hydrolysis is indicated in green, acidogenesis in orange, acetogenesis in blue and methanogenesis in yellow. A, B, C, D, E refer to the title rows in* ***Table 2****; F, G refer to the title rows in* ***Table 3****.*

biogas production and increase the share of renewable energy in total energy consumption [6–9].

Analysis of many studies on metagenomes of microbial communities from anaerobic digesters shows that (i) contribution of methanogens in the methane-yielding microbial communities is relatively small, below 20%; (ii) the most abundant phyla of bacteria are usually *Firmicutes, Bacteroidetes, Proteobacteria*, and *Actinobacteria*; (iii) methanogenic archaea are dominated by acetotrophs or hydrogenotrophs with a certain contribution of methylotrophs; (iv) substrate, operational conditions such as temperature, pH, ammonia concentration, etc. shape the structure, percentage distribution of specific taxons, and functioning of the community of microorganisms; (v) it is important to describe interactions within microbial communities and assign functions in AD steps to specific groups of microbes; and (vi) the majority of sequences are not classified at the genus level confirming that most of the microorganisms are still unrecognized [6, 10–15].

In this contribution, the purpose of the study was to prepare a list of the selected enzymes and their catalyzed reactions, being a specific enzymatic road map of AD metabolic pathways, useful in molecular studies. The available metabolic pathway databases such as KEGG PATHWAY Database [16–18], MetaCyc Metabolic Pathway Database, BioCyc Database Collection [19], and BRENDA—The Comprehensive Enzyme Information System [20] were used to select metabolic pathways dedicated only to AD from hydrolysis to methanogenic steps exerted by microbes.

2. Selected enzymes of anaerobic digestion

Figure 1 shows a scheme of AD and **Tables 1–4** present a summary of the selected enzymes and enzymatic reactions involved in decomposition of organic matter to methane and carbon dioxide. **Tables 1–4** are an extension of **Figure 1**, and in **Figure 1**, there are the links to **Tables 1–4**.

The key groups of hydrolases involved in the process of degradation of organic matter are esterases, glycosidases, and peptidases, which catalyze the cleavage of ester bonds, glycoside bonds, and peptide bonds, respectively (**Table 1**). **Table 1** also includes other classes of hydrolases such as acting on carbon-nitrogen bonds, other than peptide bonds.

In the acidogenic stage of AD, the key step is pyruvate formation from carbohydrates (**Table 2**, Part A) or other compounds and further pyruvate transformations toward short-chain fatty acids and ethanol (**Table 2**, Part B). The Part C of the **Table 2** also considers transformation of gaseous and nongaseous products of acidic fermentations, resulting from nonsyntrophic nutritional interaction between bacteria. The Parts D and E present the enzymes of glycerol and amino acid transformations, respectively. The latter requires syntrophic cooperation between microorganisms.

The enzymes catalyzing oxidation of nongaseous products of acidogenesis mainly butyrate, propionate, acetate, lactate, ethanol including the enzymes of reverse electron transfer (process responsible for energy conservation in syntrophically growing acetogens) are shown in **Table 3**.

The enzymes of the three recognized pathways of methanogenesis such as acetotrophic, hydrogenotrophic, and methylotrophic are listed in **Table 4**.

The data were prepared on the basis of detailed analysis of AD research. The enzyme nomenclature comes from the Kyoto Encyclopedia of Genes and Genomes (KEGG) database resource.

Hydrolytic enzyme	Reaction/process	EC number
Esterases	Acting on ester bonds	EC 3.1
Glycosidases	Acting on glycoside bonds	EC 3.2
	Acting on cellulose	
Cellulase; endo-1,4-beta-D-glucanase	Endohydrolysis of (1 → 4)-beta-D-glucosidic linkages in cellulose, lichenin, and cereal beta-D-glucans	EC 3.2.1.4
Cellulose 1,4-beta-cellobiosidase (nonreducing end)	Hydrolysis of (1 → 4)-beta-D-glucosidic linkages in cellulose and cellotetraose, releasing cellobiose from the nonreducing ends of the chains	EC 3.2.1.91
Beta-glucosidase	Hydrolysis of terminal, nonreducing beta-D-glucosyl residues with release of beta-D-glucose	EC 3.2.1.21
	Acting on hemicellulose	
Endo-1,4-beta-xylanase	Endohydrolysis of (1 → 4)-beta-D-xylosidic linkages in xylans	EC 3.2.1.8
Xylan 1,4-beta-xylosidase	Hydrolysis of (1 → 4)-beta-D-xylans, to remove successive D-xylose residues from the nonreducing termini	EC 3.2.1.37
Mannan endo-1,4-beta-mannosidase	Random hydrolysis of (1 → 4)-beta-D-mannosidic linkages in mannans, galactomannans, and glucomannans	EC 3.2.1.78
Beta-mannosidase	Hydrolysis of terminal, nonreducing beta-D-mannose residues in beta-D-mannosides	EC 3.2.1.25
Alpha-galactosidase	Hydrolysis of terminal, nonreducing alpha-D-galactose residues in alpha-D-galactosides, including galactose oligosaccharides, galactomannans, and galactolipids	EC 3.2.1.22
Alpha-glucuronidase	An alpha-D-glucuronoside + H_2O → an alcohol + D-glucuronate	EC 3.2.1.139
Peptidases	Acting on peptide bonds	EC 3.4
	Other hydrolases	
Hydrolases acting on carbon-nitrogen bonds, other than peptide bonds		EC 3.5
Hydrolases acting on ether bonds		EC 3.3
Hydrolases acting on carbon-carbon bonds		EC 3.7
Hydrolases acting on halide bonds		EC 3.8
Hydrolases acting on phosphorus-nitrogen bonds		EC 3.9
Hydrolases acting on sulfur-nitrogen bonds		EC 3.10
Hydrolases acting on carbon-phosphorus bonds		EC 3.11
Hydrolases acting on sulfur-sulfur bonds		EC 3.12
Hydrolases acting on carbon-sulfur bonds		EC 3.13
Hydrolases acting on acid anhydrides		EC 3.6

Table 1.
The selected enzymes of hydrolytic step of anaerobic digestion [21, 22].

Enzyme	Reaction/process	EC number
	A. Pyruvate formation from carbohydrates [23]	
	Glycolysis (the Embden-Meyerhof-Parnas pathway)	
Hexose kinase	D-Glucose + ATP ↔ D-glucose-6-phosphate + ADP	EC 2.7.1.1
Phosphoglucose isomerase	D-Glucose 6-phosphate ↔ D-fructose 6-phosphate	EC 5.3.1.9
Phosphofructose kinase	ATP + D-fructose 6-phosphate ↔ ADP + D-fructose 1,6-bisphosphate	EC 2.7.1.11
Fructose-bisphosphate aldolase	Fructose-1,6-bisphosphate ↔ dihydroxyacetone phosphate + glyceraldehyde-3-phosphate	EC 4.1.2.13
Triose phosphate isomerase	Glyceraldehyde 3-phosphate ↔ dihydroxyacetone phosphate	EC 5.3.1.1
Glyceraldehyde-3-phosphate dehydrogenase	D-Glyceraldehyde 3-phosphate + phosphate + NAD^+ ↔ 1,3-bisphosphoglycerate + NADH + H^+	EC 1.2.1.12
Phosphoglycerate kinase	1,3-Bisphosphoglycerate + ADP ↔ 3-phosphoglycerate + ATP	EC 2.7.2.3
Phosphoglycerate mutase	3-Phosphoglycerate ↔ 2-phosphoglycerate	EC 5.4.2.1
Enolase	2-Phospho-D-glycerate ↔ phosphoenolpyruvate + H_2O	EC 4.2.1.11
Pyruvate kinase	Phosphoenolpyruvate + ADP ↔ pyruvate + ATP	EC 2.7.1.40
	2-Keto-3-deoxy-6-phosphogluconate (the Entner-Doudoroff pathway)	
Glucose-6-phosphate dehydrogenase	D-glucose 6-phosphate + $NADP^+$ ↔ 6-phospho-D-glucono-1,5-lactone + NADPH + H^+	EC 1.1.1.49
Phosphogluconate dehydrogenase	6-Phospho-D-gluconate + $NAD(P)^+$ ↔ 6-phospho-2-dehydro-D-gluconate + NAD(P)H + H^+	EC 1.1.1.43
2-Keto-3-deoxy-6-phosphogluconate aldolase	2-Dehydro-3-deoxy-6-phospho-D-gluconate ↔ pyruvate + D-glyceraldehyde 3-phosphate	EC 4.1.2.14
	B. Further transformations of pyruvate—glycolytic fermentations [23–27]	
Lactate dehydrogenase	Pyruvate + NADH ↔ lactate + NAD^+	EC 1.1.1.27
Pyruvate:ferredoxin oxidoreductase, PFOR	Pyruvate + CoA + oxidized Fd ↔ acetyl-CoA + reduced Fd + CO_2 + H^+	EC 1.2.7.1
NADH:ferredoxin oxidoreductase, NFOR	Oxidized Fd + NADH ↔ reduced Fd + NAD^+ + H^+	EC 1.18.1.3
Ferredoxin hydrogenase	2 reduced ferredoxin + 2 H^+ ↔ H_2 + 2 oxidized ferredoxin	EC 1.12.7.2
Phosphotransacetylase	CoA + acetyl phosphate ↔ acetyl-CoA + phosphate	EC 2.3.1.8
Acetate kinase	ATP + acetate ↔ ADP + acetyl phosphate	EC 2.7.2.1
NAD^+-dependent ethanol dehydrogenase	Acetaldehyde + NADH + H^+ ↔ ethanol + NAD^+ An aldehyde + NADH + H^+ ↔ a primary alcohol + NAD^+	EC 1.1.1.1
Acetaldehyde dehydrogenase	Acetaldehyde + CoA + NAD^+ ↔ acetyl-CoA + NADH + H^+	EC 1.2.1.10
Acetyl-CoA acetyltransferase	2-acetyl-CoA ↔ CoA + acetoacetyl-CoA	EC 2.3.1.9
3-Hydroxybutyryl-CoA dehydrogenase	3-Acetoacetyl-CoA + NADPH + H^+ ↔ 3-hydroxybutanoyl-CoA + $NADP^+$	EC 1.1.1.157
Crotonase 3-OH-butyryl-CoA dehydratase	3-Hydroxybutanoyl-CoA ↔ crotonoyl-CoA + H_2O	EC 4.2.1.55

Enzyme	Reaction/process	EC number
2NADH+ oxidized Fd + crotonyl-CoA → 2 NAD+ reduced Fd + butyryl-CoA catalyzed by butyryl CoA dehydrogenase/electron-transfer flavoprotein complex		
Butyryl-CoA dehydrogenase	A short-chain acyl-CoA + electron-transfer flavoprotein ↔ a short-chain trans-2,3-dehydroacyl-CoA + reduced electron-transfer flavoprotein	EC 1.3.8.1
Butyryl-CoA dehydrogenase/Etf complex	Butanoyl-CoA + 2 NAD^+ + 2 reduced Fd ↔ Crotonoyl-CoA + 2 NADH + 2 oxidized Fd	EC 1.3.1.109
Phosphotransbutyrylase	Butanoyl-CoA + phosphate ↔ CoA + butanoyl phosphate	EC 2.3.1.19
Butyrate kinase	Butanoyl phosphate + ADP ↔ butanoate + ATP	EC 2.7.2.7
PFL—pyruvate formate lyase	Pyruvate + CoA ↔ acetyl-CoA + formate	EC 2.3.1.54
FHL—formate hydrogen lyase	Formate → H_2 + CO_2	EC 1.17.99.7
Pyruvate carboxylase	ATP + pyruvate + HCO_3^- ↔ ADP + phosphate + oxaloacetate	EC 6.4.1.1
Malate dehydrogenase	Malate + NAD^+ ↔ oxaloacetate + NADH + H^+	EC 1.1.1.37
Fumarate hydratase	Malate ↔ fumarate + H_2O	EC 4.2.1.2
Fumarate reductase	Fumarate + a quinol ↔ succinate + a quinone	EC 1.3.5.4
	Fumarate + NADH ↔ succinate + NAD^+	EC 1.3.1.6
Succinyl-CoA synthetase	GTP + succinate + CoA = GDP + phosphate + succinyl-CoA	EC 6.2.1.4
Methylmalonyl CoA mutase	Succinyl-CoA ↔ (R)-methylmalonyl-CoA	EC 5.4.99.2
Methylmalonyl CoA epimerase	(R)-methylmalonyl-CoA ↔ (S)-methylmalonyl-CoA	EC 5.1.99.1
Methylmalonyl-CoA decarboxylase	(S)-methylmalonyl-CoA ↔ propanoyl-CoA + CO_2	EC 4.1.1.41
Propionate-CoA transferase	Acetate + propanoyl-CoA ↔ acetyl-CoA + propanoate	EC 2.8.3.1
C. Transformation of gaseous and nongaseous products of acidic fermentations (the selected examples)		
Transformation of lactate and acetate to butyrate, hydrogen, and carbon dioxide **([28] and cited therein)**		
Lactate dehydrogenases	(S)-lactate + NAD^+ ↔ pyruvate + NADH + H^+	EC 1.1.1.27
	(R)-lactate + NAD^+ ↔ pyruvate + NADH + H^+	EC 1.1.1.28
	Lactate + 2 NAD^+ + 2 reduced Fd ↔ pyruvate + 2 NADH + 2 oxidized Fd See **Table 3**	EC 1.3.1.110
Pyruvate is oxidized to acetyl coenzyme A, which is further routed to acetate and butyrate with hydrogen release. See Part B: Further transformations of pyruvate—glycolytic fermentations		
Transformation of ethanol and acetate to butyrate and hydrogen in Clostridium kluyveri **[29]**		
Acetate kinase	See Part B. Further transformations of pyruvate—glycolytic fermentations	EC 2.7.2.1
Acetyl-CoA acetyltransferase		EC 2.3.1.9
3-Hydroxybutyryl-CoA dehydrogenase		EC 1.1.1.157
3-Hydroxyacyl-CoA dehydratase		EC 4.2.1.55
Butyryl-CoA dehydrogenase/Etf complex		EC 1.3.1.109

Enzyme	Reaction/process	EC number
Acetate CoA-transferase	Acyl-CoA + acetate ↔ a fatty acid anion + acetyl-CoA	EC 2.8.3.8
***Reductive carbon monoxide dehydrogenase/acetyl-CoA synthase pathway (reductive CODH/ACS)* [30]**		
NADP-dependent formate dehydrogenase	CO_2 + NADPH ↔ formate + $NADP^+$	EC 1.17.1.10
Formyltetrahydrofolate synthetase	ATP + formate + tetrahydrofolate ↔ ADP + phosphate + 10-formyltetrahydrofolate	EC 6.3.4.3
Methenyltetrahydrofolate cyclohydrolase	10-Formyltetrahydrofolate ↔ 5,10-methenyltetrahydrofolate + H_2O	EC 3.5.4.9
NADP-dependent methylenetetrahydrofolate dehydrogenase	5,10-Methenyltetrahydrofolate + NADPH + H^+ ↔ 5,10-Methylenetetrahydrofolate + $NADP^+$	EC 1.5.1.5
Ferredoxin-dependent methylenetetrahydrofolate reductase	5,10-Methylenetetrahydrofolate + 2 reduced Fd + 2 H^+ ↔ 5-methyltetrahydrofolate + 2 oxidized Fd	EC 1.5.7.1
5,10-Methylenetetrahydrofolate reductase	5,10-Methylenetetrahydrofolate + NAD(P)H + H^+ ↔ 5-methyltetrahydrofolate + $NAD(P)^+$	EC 1.5.1.20
5-Methyltetrahydrofolate: corrinoid/iron–sulfur protein Co-methyltransferase	[Co(I) corrinoid Fe-S protein] + 5-methyltetrahydrofolate ↔ [methyl-Co(III) corrinoid Fe-S protein] + tetrahydrofolate	EC 2.1.1.258
Carbon monoxide dehydrogenase	CO_2 + 2 reduced Fd + 2 H^+ ↔ CO + H_2O + 2 oxidized Fd	EC 1.2.7.4
CO-methylating acetyl-CoA synthase	CO + CoA + [methyl-Co(III) corrinoid Fe-S protein] ↔ acetyl-CoA + [Co(I) corrinoid Fe-S protein]	EC 2.3.1.169
D. Glycerol transformations [31, 32]		
Oxidative pathway		
Glycerol dehydrogenase	Glycerol + NAD^+ ↔ glycerone (dihydroxyacetone) + NADH + H^+	EC 1.1.1.6
Dihydroxyacetone kinase	ATP + glycerone ↔ ADP + glycerone phosphate	EC 2.7.1.29
For further reactions, see Part A: Pyruvate formation		
Reductive pathway		
Glycerol dehydratase	Glycerol ↔ 3-hydroxypropionaldehyde + H_2O	EC 4.2.1.30
1,3-Propanediol dehydrogenase	3-Hydroxypropionaldehyde + NADH + H^+ ↔ 1,3-propanediol + NAD^+	EC 1.1.1.202
E. Amino acids fermentations [33–37]		
Syntrophy with H_2-scavenging microorganism: amino acid degradation involves NAD(P)- or FAD-dependent deamination of amino acids to the corresponding α-keto acids by amino acid dehydrogenases **(EC 1.4.1.X)**: $RCH(NH_4^+)COO^- + H_2O \rightarrow RCOCOO^- + NH_4^+ + H_2$ and further conversion of α-keto acids via oxidative decarboxylation to fatty acids: $RCOCOO^- + H_2O \rightarrow RCOO^- + CO_2 + H_2$ [33]		
Without syntrophy with H_2-scavenging microorganism: **Stickland Reaction**—coupled oxidation-reduction reactions between suitable amino acids (coupled deamination of amino acids); one member of the pair is oxidized (dehydrogenated) and the other is reduced (hydrogenated) [34], for example, Alanine and glycine: alanine + 2 glycine + $3H_2O \rightarrow$ 3 acetate$^-$ + $3NH_4^+ + HCO_3^- + H^+$ Valine and glycine: valine + 2 glycine + $3H_2O \rightarrow$ isobutyrate$^-$ + 2 acetate$^-$ + $3NH_4^+ + HCO_3^- + H^+$ Leucine and glycine: leucine + 2 glycine + $3H_2O \rightarrow$ isovalerate$^-$ + 2 acetate$^-$ + $3NH_4^+ + HCO_3^- + H^+$		

Enzyme	Reaction/process	EC number
Examples of amino acid dehydrogenases catalyzing deamination of amino acids to the corresponding α-keto acids [33]		
Aspartate dehydrogenase	L-aspartate + H_2O + $NAD(P)^+$ ↔ oxaloacetate + NH_3 + NAD(P)H + H^+	EC 1.4.1.21
Valine dehydrogenase	L-valine + H_2O + $NADP^+$ ↔ 3-methyl-2-oxobutanoate + NH_3 + NADPH + H^+	EC 1.4.1.8
Alanine dehydrogenase	L-alanine + H_2O + NAD^+ ↔ pyruvate + NH_3 + NADH + H^+	EC 1.4.1.1
Leucine dehydrogenase	L-leucine + H_2O + NAD+ ↔ 4-methyl-2-oxopentanoate + NH_3 + NADH + H^+	EC 1.4.1.9
Key enzymes of Stickland reaction [34–36]		
Glycine reductase GR pathway (*grd* operon)		
Glycine reductase	Glycine + phosphate + reduced thioredoxin + H^+ ↔ acetyl phosphate + NH_3 + oxidized thioredoxin + H_2O	EC 1.21.4.2
Acetate kinase	Acetyl phosphate + ADP ↔ acetate + ATP	EC 2.7.2.1
Proline reductase PR pathway (*prd* operon)		
D-proline reductase (dithiol)	D-proline + dihydrolipoate ↔5-aminopentanoate (5-aminovalerate) + lipoate	EC 1.21.4.1
Others examples [33]		
Serine dehydratase	L-serine ↔ pyruvate + NH_3 (overall reaction) (1a) L-serine ↔ 2-aminoprop-2-enoate + H_2O (1b) 2-Aminoprop-2-enoate ↔ 2-iminopropanoate (spontaneous) (1c) 2-Iminopropanoate + H_2O ↔ pyruvate + NH_3 (spontaneous)	EC 4.3.1.17
Threonine dehydratase	L-threonine ↔ 2-oxobutanoate + NH_3 (overall reaction) (1a) L-threonine ↔ 2-aminobut-2-enoate + H_2O; (1b) 2-Aminobut-2-enoate ↔ 2-iminobutanoate (spontaneous) (1c) 2-Iminobutanoate + H_2O ↔ 2-oxobutanoate + NH_3 (spontaneous)	EC 4.3.1.19
Detailed pathways of glutamate fermentation via 3-methylaspartate [37]		
Glutamate mutase (methylaspartate mutase)	L-glutamate ↔ L-threo-3-methylaspartate	**EC 5.4.99.1**
Methyl aspartase	L-threo-3-methylaspartate ↔ mesaconate (2-methylfumarate) + NH_3	EC 4.3.1.2
Mesaconase (2-methylmalate dehydratase)	2-Methylfumarate + H2O ↔ (S)-2-methylmalate	4.2.1.34
Citramalate lyase	(2S)-2-hydroxy-2-methylbutanedioate ↔ acetate + pyruvate (S)-2-methylmalate = 2-hydroxy-2-methylbutanedioate	4.1.3.22
For further transformations of pyruvate to acetate and butyrate, see Part B.		
For further transformations of pyruvate to propionate, see Part B.		
Detailed pathway of glutamate fermentation via 2-hydroxyglutarate [37]		
Glutamate dehydrogenase	L-glutamate + H_2O + NAD^+ ↔ 2-oxoglutarate + NH_3 + NADH + H^+	1.4.1.2

Enzyme	Reaction/process	EC number
2-Hydroxyglutarate dehydrogenase	(S)-2-hydroxyglutarate + acceptor ↔ 2-oxoglutarate + reduced acceptor	1.1.99.2
Glutaconate (2-hydroxyglutarate) CoA-transferase	Acetyl-CoA + (E)-glutaconate ↔ acetate + glutaconyl-1-CoA	2.8.3.12
2-Hydroxyglutaryl-CoA dehydratase	**(R)-2-hydroxyglutaryl-CoA ↔ (E)-glutaconyl-CoA + H_2O**	**EC 4.2.1.167**
Glutaconyl-CoA decarboxylase	4-Carboxybut-2-enoyl-CoA ↔ but-2-enoyl-CoA + CO_2	4.1.1.70

Table 2.
*The selected enzymes of acidogenic step of anaerobic digestion. A, B, C, D, and E refer to the processes indicated in **Figure 1**.*

Enzyme	Reaction/process	EC number
F. Acetogenesis dependent on syntrophic relations between microorganisms		
Acetate oxidation by, for example, *Clostridium ultunense*—oxidative carbon monoxide dehydrogenase/acetyl-CoA synthase pathway (oxidative CODH/ACS): Acetate$^-$ + $4H_2O \rightarrow 2\ HCO_3^-$ + $4H_2$ + H^+, $\Delta G^{0'}$ = + 104.6 kJ/mol, with the H_2 consuming methanogen, $\Delta G^{0'}$ = −31.0 kJ/mol [38]		
NADP-dependent formate dehydrogenase		See **Table 2**, Part C
Formyltetrahydrofolate synthetase		
Methenyltetrahydrofolate cyclohydrolase		
NADP-dependent methylenetetrahydrofolate dehydrogenase		
Ferredoxin-dependent methylenetetrahydrofolate reductase		
5,10-Methylenetetrahydrofolate reductase		
5-Methyltetrahydrofolate:corrinoid/iron-sulfur protein Co-methyltransferase		
Carbon monoxide dehydrogenase		
CO-methylating acetyl-CoA synthase		
Reverse electron transfer during acetate oxidation has yet to be confirmed. Direct interspecies electron transfer (DIET) is not excluded (Westerholm et al., 2016)		
Acetate oxidation by *Geobacter sulfurreducens*: Acetate oxidation coupled to reduction of fumarate to succinate ($\Delta G^{0'}$ = −249 kJ per mol acetate), acetate metabolism proceeds via reactions of the citric acid cycle [39]		
Acetate kinase		See **Table 2**, Part B
Phosphotransacetylase		
Citric acid cycle		
Citrate synthase	Acetyl-CoA + H_2O + oxaloacetate ↔ citrate + CoA	EC 2.3.3.1
Aconitase	Citrate ↔ isocitrate (overall reaction)	EC 4.2.1.3
Isocitrate dehydrogenase ($NADP^+$-dependent)	Isocitrate + $NADP^+$ ↔ 2-oxoglutarate + CO_2 + NADPH + H^+	EC1.1.1.42
2-Oxoglutarate:ferredoxin oxidoreductase	2-Oxoglutarate + CoA + 2 oxidized Fd = succinyl-CoA + CO_2 + 2 reduced Fd + 2 H^+	EC 1.2.7.3
Succinyl-CoA:acetate CoA-transferase	Succinyl-CoA + acetate ↔ acetyl-CoA + succinate	EC 2.8.3.18

Enzyme	Reaction/process	EC number
Succinate dehydrogenase	succinate + a quinone ↔ fumarate + a quinol	EC 1.3.5.1
Fumarate hydratase	(S)-malate ↔ fumarate + H_2O	EC 4.2.1.2
Malate dehydrogenase	(S)-malate + NAD^+ ↔ oxaloacetate + NADH + H^+	EC 1.1.1.37
Butyrate oxidation by *Syntrophomonas wolfei*: Butyrate$^-$ + $2H_2O$ → 2 acetate$^-$ + $2H^+$ + $2H_2$, $\Delta G^{0'}$ = + 48.3 kJ/mol, with the H_2 consuming methanogen, $\Delta G^{0'}$ = −17.3 kJ/mol [4]		
CoA transferase	Butyrate + acetyl-CoA ↔ butyryl-CoA + acetate	EC 2.8.3.9
Butyryl-CoA dehydrogenase		See **Table 2**, Part B
Crotonase-3-OH-butyryl-CoA dehydratase		
3-Acetyl-CoA acetyltransferase		
Hydroxybutyryl-CoA dehydrogenase		
Phosphotransacetylase		
Acetate kinase		
Butyrate oxidation coupled with a reverse electron transfer that involves electron transfer flavoprotein EtfAB, membrane-anchored electron carrier DUF224 protein, the menaquinone pool in the membrane, a membrane-bound cytochrome, NADH:hydrogenase/formate-dehydrogenase complex (NDH/HYD1/FDH-1 complex), Rnf (proton-translocating ferredoxin:NAD^+ oxidoreductase) [40]		
Propionate oxidation by *Syntrophobacter wolinii*: Propionate$^-$ + $3H_2O$ → acetate$^-$ + HCO_3^- + H^+ + $3H_2$, $\Delta G^{0'}$ = + 76.0 kJ/mol, with the H_2 consuming methanogen, $\Delta G^{0'}$ = −22.4 kJ/mol [4]		
Pyruvate carboxylase		See **Table 2**, Part B
Malate dehydrogenase		
Fumarate hydratase		
Fumarate reductase		
Succinate dehydrogenase	Succinate + a quinone ↔ fumarate + a quinol	EC 1.3.5.1
Succinyl-CoA synthetase		See **Table 2**, Part B
Methylmalonyl CoA mutase		
Methylmalonyl CoA epimerase		
Methylmalonyl-CoA decarboxylase		
Propionate-CoA transferase		
Propionate oxidation coupled with a reverse electron transfer that involves menaquinone, proteins encoded by cytochrome c homologous genes, cytochrome b:quinone oxidoreductases, formate dehydrogenases, hydrogenases including confurcating [FeFe]-hydrogenases [41]		
Six syntrophy-specific functional domains found in the genomes of the butyrate- or propionate-oxidizing syntrophs [42]	**InterPro number**	
Extra-cytoplasmic formate dehydrogenase (FDH) alpha subunit, EC 1.17.1.9	IPR006443	
FdhE-like protein—tightly connected with FDH	IPR024064	

Enzyme	Reaction/process	EC number
FDH accessory protein—tightly connected with FDH	IPR006452	
CapA—a membrane-bound complex, a protein involved in capsule or biofilm formation that may facilitate syntrophic growth (*also present in acetate-oxidizers*)	IPR019079	
FtsW, RodA, SpoVE—membrane-integrated proteins involved in membrane integration, cell division, sporulation, and shape determination	IPR018365	
Ribonuclease P involved in tRNA maturation	IPR020539	
Functional domains involved in electron transfer identified by [42]	**InterPro number**	
Cytoplasmic FDH	IPR027467, IPR006655, IPR006478, IPR019575, IPR001949	
Extracytoplasmic FDH	IPR006443	
Formate transporter	IPR000292, IPR024002	
Fe-Fe hydrogenase	IPR004108, IPR009016, IPR003149, IPR013352	
NiFe hydrogenase	IPR001501, IPR018194	
Rnf complex: 2 reduced Fd + NAD^+ + H^+ + Na^+ ↔ 2 oxidized Fd + NADH + Na^+ (EC 1.18.1.8)	IPR007202, IPR010207, IPR026902, IPR010208, IPR004338, IPR011303, IPR007329	
Ech complex: 2 reduced Fd + $NADP^+$ + H^+ ↔ 2 oxidized Fd + NADPH (EC 1.18.1.2)	IPR001750, IPR001516, IPR001694, IPR006137, IPR001268, IPR012179, IPR001135	
Etf alpha, Etf beta, Bcd (Butyryl-CoA dehydrogenase): see **Table 2**, Part B (EC 1.3.1.109)	IPR014731, IPR012255, IPR006089, IPR009075, IPR006092, IPR006091, IPR013786, IPR009100	
Cytochromes: c cIII b561 b5	IPR023155, IPR024673 IPR020942, IPR002322 IPR016174, IPR000516 IPR001199	
DUF224 protein complex	IPR003816, IPR004017, IPR023234	
Lactate oxidation by *Desulfovibrio vulgaris:* $Lactate^- + H_2O \rightarrow acetate^- + CO_2 + 4\ H_2$, $\Delta G^{0'} = -8.8$ kJ/mol with the H_2 consuming methanogen, $\Delta G^{0'} = -74.2$ kJ/mol [43]		
	Lactate dehydrogenase	See **Table 2**, Part B
	Pyruvate:ferredoxin oxidoreductase	
	Phosphate acetyltransferase	
	Acetate kinase	
	Alcohol dehydrogenase	
Lactate oxidation coupled with a reverse electron transfer that involves the membrane-bound Qmo complex, cytochromes, hydrogenases (Coo, Hyn, Hyd, Hys), formate dehydrogenases, menaquinone, membrane-bound Qrc complex [43, 44]		

Enzyme	Reaction/process	EC number
Ethanol oxidation by *Pelobacter carbinolicus* Ethanol + $H_2O \rightarrow$ acetate$^-$ + H^+ + $2H_2$, $\Delta G^{0'} = +9.6$ kJ/mol with the H_2 consuming methanogen, $\Delta G^{0'} = -56$ kJ/mol [4]		
NAD$^+$-dependent ethanol dehydrogenase		See **Table 2**, Part B
Acetaldehyde dehydrogenase (acetylating)		
Nonacetylating acetaldehyde dehydrogenase	An aldehyde + NAD$^+$ + $H_2O \leftrightarrow$ a carboxylate + NADH + H$^+$	EC 1.2.1.3
Phosphotransacetylase		See **Table 2**, Part B
Acetate kinase		
Ethanol oxidation coupled with a reverse electron transfer that involves membrane-bound ion-translocating ferredoxin:NAD$^+$ oxidoreductase, formate dehydrogenases, and confurcating hydrogenases [1, 45]		
G. Acetogenesis independent on syntrophic relations between microorganisms		
Ethanol oxidation by *Acetobacterium woodii*: 2 ethanol + 2 $CO_2 \rightarrow$ 3 acetate—75.4 kJ/mol [46]		
Bifunctional acetaldehyde-CoA/alcohol dehydrogenase	Ethanol + NAD$^+$ $\rightarrow$ acetaldehyde + NADH + H$^+$ acetaldehyde + NAD$^+$ + CoA $\rightarrow$ acetyl-CoA + 2 NADH + H$^+$ Ethanol is oxidized to acetyl-CoA in a two-step reaction by a bifunctional acetylating ethanol/aldehyde dehydrogenase	[EC:1.2.1.10 1.1.1.1]
Acetyl-CoA is transformed to acetate with the release of ATP		See **Table 2**, Part B
Reduction of ferredoxin by NADH by reverse electron flow in a reaction catalyzed by Rnf complex		See Part F
Carbon dioxide is reduced to acetate via the Wood-Ljungdahl pathway		See **Table 2**, Part C
Lactate oxidation by *Acetobacterium woodii*: 2 lactate $\rightarrow$ 3 acetate—61 kJ/mol [47]		
Lactate dehydrogenase	Lactate + 2 NAD$^+$ + 2 reduced Fd $\leftrightarrow$ pyruvate + 2 NADH + 2 oxidized Fd The enzyme uses flavin-based electron confurcation to drive endergonic lactate oxidation with NAD$^+$ as oxidant at the expense of simultaneous exergonic electron flow from reduced ferredoxin to NAD$^+$	EC 1.3.1.110
Pyruvate is transformed to acetyl-CoA and further to acetate with the release of ATP		See **Table 2**, Part B
Reduction of ferredoxin by NADH by reverse electron flow in a reaction catalyzed by Rnf complex		See Part F
Carbon dioxide is reduced to acetate via the Wood-Ljungdahl pathway		See **Table 2**, Part C

Table 3.
The selected enzymes of acetogenic step of anaerobic digestion. F and G refer to the processes indicated in ***Figure 1****.*

Enzyme	Reaction/process	EC number
MFR—methanofuran, H-S-CoM—coenzyme M, H-S-CoB—coenzyme B, H_4MPT—tetrahydromethanopterin, F_{420}—5'deazaflavin, H_4SPT—tetrahydrosarcinapterin		
Hydrogenotrophic pathway		
Formylmethanofuran dehydrogenase	CO_2 + MFR + 2 reduced Fd + $2H^+$ ↔ formyl-MFR + H_2O + 2 oxidized Fd	EC 1.2.7.12
Formylmethanofuran-H_4MPT formyltransferase	Formyl-MFR + H_4MPT ↔ MFR + formyl-H_4MPT	EC 2.3.1.101
Methenyl-H_4MPT cyclohydrolase	Formyl-H_4MPT + H^+ ↔ methenyl-H_4MPT + H_2O	EC 3.5.4.27
F_{420}-dependent methylene-H_4MPT dehydrogenase	Methenyl-H_4MPT + reduced F_{420} ↔ methylene-H_4MPT + oxidized F_{420}	EC 1.5.98.1
H_2-forming methylene-H_4MPT dehydrogenase	Methenyl-H_4MPT + H_2 ↔ methylene-H_4MPT + H^+	EC 1.12.98.2
F_{420}-dependent methylene-H_4MPT reductase	Methylene-H_4MPT + reduced F_{420} ↔ CH_3-H_4MPT + oxidized F_{420}	EC 1.5.98.2
Methyl-H_4MPT:coenzyme M methyl-transferase	Coenzyme M + methyl-H_4MPT + 2 Na+/in ↔ 2-methyl-coenzyme M + 2 Na+/out + H_4MPT	EC 2.1.1.86
Methyl-CoM reductase	CH_3-S-CoM + H-S-CoB ↔ CoM-S-S-CoB + CH_4	EC 2.8.4.1
Heterodisulfide reductase	CoM-S-S-CoB + dihydromethanophenazine ↔ CoB + CoM + methanophenazine	EC 1.8.98.1
Acetotrophic pathway		
Acetate kinase-phosphotransacetylase system in *Methanosarcina*; acetate thiokinase in *Methanosaeta*	Acetate + CoA ↔ acetyl-CoA + H_2O	EC 2.7.2.1 EC 2.3.1.8 EC 6.2.1.1
CO-methylating acetyl-CoA synthase	Acetyl-CoA + a [Co(I) corrinoid Fe-S protein] ↔ CO + CoA + [methyl-Co(III) corrinoid Fe-S protein]	EC 2.3.1.169
5-Methyltetrahydrosarcinapterin: corrinoid/iron-sulfur protein Co-methyltransferase	[Methyl-Co(III) corrinoid Fe-S protein] + tetrahydrosarcinapterin ↔ a [Co (I) corrinoid Fe-S protein] + 5-methyltetrahydrosarcinapterin	EC 2.1.1.245
Anaerobic carbon monoxide dehydrogenase	CO + H_2O + 2 oxidized Fd ↔ CO_2 + 2 reduced Fd + 2 H^+	EC 1.2.7.4
Methyl H_4SPT: coenzyme M methyltransferase	CH_3 H_4SPT + H-S-CoM ↔ CH_3-S-CoM + H_4SPT	EC 2.1.1.-
Methyl-CoM reductase	CH_3-S-CoM + H-S-CoB ↔ CoM-S-S-CoB + CH_4	EC 2.8.4.1
Heterodisulfide reductase	CoM-S-S-CoB + dihydromethanophenazine ↔ CoB + CoM + methanophenazine	EC 1.8.98.1
Methylotrophic pathway		
Methanol:corrinoid protein Co-methyltransferase	Methanol + Co(I) corrinoid protein ↔ Methyl-Co(III) corrinoid protein + H_2O	EC 2.1.1.90
[Methyl-Co(III) corrinoid protein]: coenzyme M methyltransferase	Coenzyme M + Methyl-Co(III) corrinoid protein ↔ 2-(methylthio)ethanesulfonate + Co(I) corrinoid protein	EC 2.1.1.246

Enzyme	Reaction/process	EC number
Methylamine:corrinoid protein Co-methyltransferase	Methylamine + [Co(I) methylamine-specific corrinoid protein] ↔ a [methyl-Co(III) methylamine-specific corrinoid protein] + NH_3	EC 2.1.1.248
Dimethylamine:corrinoid protein Co-methyltransferase	Dimethylamine + [Co(I) dimethylamine-specific corrinoid protein] ↔ a [methyl-Co(III) dimethylamine-specific corrinoid protein] + methylamine	EC 2.1.1.249
Trimethylamine:corrinoid protein Co-methyltransferase	Trimethylamine + a [Co(I) trimethylamine-specific corrinoid protein] ↔ a [methyl-Co(III) trimethylamine-specific corrinoid protein] + dimethylamine	EC 2.1.1.249
[Methyl-Co(III) methylamine-specific corrinoid protein]:coenzyme M methyltransferase	[Methyl-Co(III) methylamine-specific corrinoid protein] + CoM ↔ methyl-CoM + a [Co(I) methylamine-specific corrinoid protein]	EC 2.1.1.247
Methyl-CoM reductase	CH_3-S-CoM + H-S-CoB ↔ CoM-S-S-CoB + CH_4	EC 2.8.4.1
Heterodisulfide reductase	CoM-S-S-CoB + dihydromethanophenazine ↔ CoB + CoM + methanophenazine	EC 1.8.98.1

Table 4.
The selected enzymes of methanogenic step of anaerobic digestion [48, 49].

3. Conclusion

Biomass conversion to methane and carbon dioxide is the effect of complex interactions between microorganisms. These processes occur due to the microbial enzymatic machinery involved in specific metabolic pathways. Meta-omic analyses of microbial communities involved in AD reveal (i) dependence of microbial communities on the type of feedstock and operational conditions and (ii) describe interactions within microbial communities and ecophysiological functions of the specific taxa. Searching for the gene presence, gene expression, and protein expression, as well as linking structure and function of microbial communities, allows to develop a fundamental understanding of AD. This chapter is believed to contribute to the studies on the enzymatic road map of anaerobic digestion. However, it is only the tip of the iceberg of processes occurring in the microbial cells/microbial communities.

Acknowledgements

We acknowledge the support of The National Science Centre, Poland, through grant UMO-2015/17/B/NZ9/01718 and The National Centre for Research and Development, Poland, through grant BIOSTRATEG2/297310/13/NCBiR/2016.

Conflict of interest

The authors declare that there are no conflicts of interest.

Author details

Anna Sikora[1]*, Anna Detman[1], Damian Mielecki[1], Aleksandra Chojnacka[1] and Mieczysław Błaszczyk[2]

1 Institute of Biochemistry and Biophysics—Polish Academy of Sciences, Warsaw, Poland

2 Faculty of Agriculture and Biology, Warsaw University of Life Sciences, Warsaw, Poland

*Address all correspondence to: annaw@ibb.waw.pl

References

[1] Sieber JR, McInerney MJ, Gunsalus RP. Genomic insights into syntrophy: The paradigm for anaerobic metabolic cooperation. Annual Review Microbiology. 2012;**66**:429-452. DOI: 10.1146/annurev-micro-090110-102844

[2] Mao CL, Feng YZ, Wang XJ, Ren GX. Review on research achievements of biogas from anaerobic digestion. Renewable and Sustinable Energy Reviews. 2015;**45**:540-555. DOI: 10.1016/j.rser.2015.02.032

[3] Sikora A, Detman A, Chojnacka A, Błaszczyk MK. Anaerobic digestion: I. A common process ensuring energy flow and the circulation of matter in ecosystems. II. A tool for the production of gaseous biofuels. In: Jozala AF, editor. Fermentation Processes. Rijeka: InTech; 2017. pp. 271-301. DOI: 10.5772/64645

[4] Kamagata Y. Syntrophy in anaerobic digestion. In: Fang HP, Zhang T, editors. Anaerobic Biotechnology: Environmental Protection and Resource Recovery. London: Imperial College Press, World Scientific; 2015. pp. 13-32. DOI: 10.1142/p1034/suppl_file/p1034_chap02

[5] Stams AJM, Plugge CM. Electron transfer in syntrophic communities of anaerobic bacteria and archaea. Nature Reviews. 2009;7:568-577. DOI: 10.1038/nrmicro2166

[6] Kleinsteuber S. Metagenomics of methanogenic communities in anaerobic digesters. Biogenesis of hydrocarbons. In: Stams AJM, Sousa DZ, editors. Biogenesis of Hydrocarbons, Handbook of Hydrocarbon and Lipid Microbiology. Springer Nature Switzerland: Springer International Publishing AG; 2018. pp. 1-23. DOI: 10.1007/978-3-319-53114-4_16-1

[7] Vanwonterghem I, Jensen PD, Ho DP, Batstone DJ, Tyson GW. Linking microbial community structure, interactions and function in anaerobic digesters using new molecular techniques. Current Opinion Biotechnology. 2014;**27**:55-64. DOI: 10.1016/j.copbio.2013.11.004

[8] Koch C, Müller S, Harms H, Harnisch F. Microbiomes in bioenergy production: From analysis to management. Current Opinion Biotechnology. 2014;**27**:65-72. DOI: 10.1016/j.copbio.2013.11.006

[9] Abram F. Systems-based approaches to unravel multi-species mirobial community functioning. Computational and Structural Biotechnology Journal. 2015;**13**:24-32. DOI: 10.1016/j.csbj.2014.11.009

[10] Cai M, Wilkins D, Chen J, Ng S-K, Lu H, Jia Y, et al. Metagenomic reconstruction of key anaerobic digestion pathways in municipal sludge and industrial wastewater biogas-producing systems. Frontiers in Microbiology. 2016;7:778. DOI: 10.3389/fmicb.2016.00778

[11] Granada CE, Hasan C, Marder M, Konrad O, Vargas LK, Passaglia LMP, et al. Biogas from slaughterhouse wastewater anaerobic digestion is driven by the archaeal family *Methanobacteriaceae* and bacterial families *Porphyromonadaceae* and *Tissierellaceae*. Renewable Energy. 2018; **118**:840-846. DOI: 10.1016/j.renene.2017.11.077

[12] Delforno TP, Lacerda GV Jr, Sierra-Garcia IN, Okada DY, Macedo TZ, Varesche MBA, et al. Metagenomic analysis of the microbiome in three different bioreactor configurations applied to commercial laundry wastewater treatment. Science of the

Total Environment. 2017;**587-588**: 389-398. DOI: 10.1016/j.scitotenv.2017.02.170

[13] Campanaro S, Treu L, Kougias PG, De Francisci D, Valle G, Angelidaki I. Metagenomic analysis and functional characterization of the biogas microbiome using high throughput shotgun sequencing and a novel binning strategy. Biotechnology for Biofuels. 2016;**9**:26. DOI: 10.1186/s13068-016-0441-1

[14] Luo G, Fotidis IA, Angelidaki I. Comparative analysis of taxonomic, functional, and metabolic patterns of microbiomes from 14 full-scale biogas reactors by metagenomic sequencing and radioisotopic analysis. Biotechnology for Biofuels. 2016;**9**:51. DOI: 10.1186/s13068-016-0465-6

[15] Guo J, Peng Y, Ni B-J, Han X, Fan L, Yuan Z. Dissecting microbial community structure and methane-producing pathways of a full-scale anaerobic reactor digesting activated sludge from wastewater treatment by metagenomics sequencing. Microbial Cell Factories. 2015;**14**:33. DOI: 10.1186/s12934-015-0218-4

[16] Kanehisa M, Furumichi M, Tanabe M, Sato Y, Morishima K. KEGG: New perspectives on genomes, pathways, diseases and drugs. Nucleic Acids Research. 2017;**45**:D353-D361. DOI: 10.1093/nar/gkw1092

[17] Kanehisa M, Sato Y, Kawashima M, Furumichi M, Tanabe M. KEGG as a reference resource for gene and protein annotation. Nucleic Acids Research. 2016;**44**:D457-D462. DOI: 10.1093/nar/gkv1070

[18] Kanehisa M, KEGG GS. Kyoto Encyclopedia of genes and genomes. Nucleic Acids Research. 2000;**28**:27-30. DOI: 10.1093/nar/27.1.29

[19] Caspi R, Billington R, Fulcher CA, Keseler IM, Kothari A, et al. The MetaCyc database of metabolic pathways and enzymes. Nucleic Acids Research. 2018;**46**(Database issue): D633-D639 http://doi.org/10.1093/nar/gkx935

[20] BRENDA—The Comprehensive Enzyme Information System. Available from: https://www.brenda-enzymes.org/. [Accessed: July 24, 2018]

[21] The Enzyme List Class 3—Hydrolases. Nomenclature Committee of the International Union of Biochemistry and Molecular Biology (NC-IUBMB) Generated from the ExplorEnz database; 2010

[22] Shresthaa S, Fonolla X, Khanal SK, Raskina L. Biological strategies for enhanced hydrolysis of lignocellulosic biomass during anaerobic digestion: Current status and future perspectives. Bioresource Technology. 2017;**245**: 1245-1257. DOI: 10.1016/j.biortech.2017.08.089

[23] Berg JM, Tymoczko JL, Gatto GJ, Stryer L. Biochemistry. 8th ed. New York: W.H. Freeman & Company; 2015

[24] Angenent LT, Karim K, Al-Dahhan MH, Wrenn BA, Domiguez-Espinosa R. Production of bioenergy and biochemicals from industrial and agricultural wastewater. Trends in Biotechnology. 2004;**22**: 477-485. DOI: 10.1016/j.tibtech.2004.07.001

[25] Hallenbeck PC. Fundamentals of the fermentative production of hydrogen. Water Science & Technology. 2005;**52**: 21-29. PMID: 16180405

[26] Kraemer JT, Bagley DM. Improving the yield from fermentative hydrogen production. Biotechnology Letters. 2007;**29**:685-695. DOI: 10.1007/s10529-006-9299-9

[27] Lee D-J, Show K-Y, Su A. Dark fermentation on biohydrogen production: Pure cultures. Bioresource Technology. 2011;**102**(18):8393-8402

[28] Sikora A, Błaszczyk M, Jurkowski M, Zielenkiewicz U. Lactic acid bacteria in hydrogen producing consortia: On purpose or by coincidence. In: Kongo M, editor. Lactic Acid Bacteria—R & D for Food, Health and Livestock Purposes. Rijeka, InTech; 2013. pp. 487-514. DOI: 5772/50364

[29] Li F, Hinderberger J, Seedorf H, Zhang J, Buckel W, Thauer RK. Coupled ferredoxin and crotonyl coenzyme A (CoA) reduction with NADH catalyzed by the butyryl-CoA dehydrogenase/Etf complex from *Clostridium klyuveri*. Journal of Bacteriology. 2008;**190**: 843-850. DOI: 10.1128/JB.01417-07

[30] Diekert G, Wohlfarth G. Metabolism of homoacetogens. Antonie Van Leeuwenhoek. 1994;**66**:209-221. DOI: 10.1007/BF00871640

[31] Viana MB, Freitas AV, Leitão RC, Pinto GAS, Santaella ST. Anaerobic digestion of crude glycerol: A review. Environmental Technology Reviews. 2012;**1**:81-92. DOI: 10.1080/09593330.2012.692723

[32] Ganzle MG. Lactic metabolism revisited: Metabolism of lactic acid bacteria in food fermentations and food spoilage. Current Opinion in Food Science. 2015;**2**:106-117. DOI: 10.1016/j.cofs.2015.03.001

[33] Schink B, Stams AJM. Syntrophism among prokaryotes. In: Dworkin M, editor. The Prokaryotes. 3rd ed. New York: Springer; 2006. pp. 309-335. DOI: 10.1007/978-3-642-30123-0_59

[34] Nisman B. The Stickland reaction. Bacteriology Reviews. 1954;**18**:16-42

[35] Bouillaut L, Self WT, Sonensheina AL. Proline-dependent regulation of *Clostridium difficile* Stickland metabolism. Journal of Bacteriology. 2013;**195**:844-854. DOI: 10.1128/JB.01492-12

[36] Fonknechten N, Chaussonnerie S, Tricot S, Lajus A, Andreesen JR, et al. *Clostridium sticklandii*, a specialist in amino acid degradation: Revisiting its metabolism through its genome sequence. BMC Genomics. 2010;**11**:555. DOI: 10.1186/1471-2164-11-555

[37] Buckel W. Unusual enzymes involved in five pathways of glutamate fermentation. Applied Microbiology and Biotechnology. 2001;**57**:263-273. DOI: 10.1007/s002530100773

[38] Hattori S. Syntrophic acetate-oxidizing microbes in methanogenic environments. Microbes and Environments. 2008;**23**:118-127. DOI: 10.1264/jsme2.23.118

[39] Galushko AS, Schink B. Oxidation of acetate through reactions of the citric acid cycle by *Geobacter sulfurreducens* in pure culture and in syntrophic coculture. Archives of Microbiology. 2000;**174**:314-321. DOI: 10.1007/s002030000208

[40] Schmidt A, Müller N, Schink B, Schleheck D. A proteomic view at the biochemistry of syntrophic butyrate oxidation in *Syntrophomonas wolfei*. PLoS One. 2013;**8**(2):e56905. DOI: 10.1371/journal.pone.0056905

[41] Muller N, Worm P, Schink B, Stams AJM, Plugge CM. Syntrophic butyrate and propionate oxidation processes: From genomes to reaction mechanisms. Environmenal Microbiology Reports. 2010;**2**:489-499. DOI: 10.1111/j.1758-2229.2010.00147.x

[42] Worm P, Koehorst JJ, Visser M, Sedano-Núñez VT, Schaap PJ, et al. A genomic view on syntrophic versus non-syntrophic lifestyle in anaerobic fatty

acid degrading communities. Biochimica et Biophysica Acta Bioenergetics. 2014;**1837**:2004-2016. DOI: 10.1016/j.bbabio.2014.06.005

[43] Walker CB, He Z, Yang ZK, Ringbauer JA, He Q, et al. The electron transfer system of syntrophically grown *Desulfovibrio vulgaris*. Journal of Bacteriology. 2009;**191**:5793-5801. DOI: 10.1128/JB.00356-09

[44] Meyer B, Kuehl J, Deutschbauer AM, Price MN, Arkin AP, et al. Variation among *Desulfovibrio* species in electron transfer systems used for syntrophic growth. Journal of Bacteriology. 2013;**195**(5):990-1004. DOI: 10.1128/JB.01959-12

[45] Schmidt A, Frensch M, Schleheck D, Schink B, Muller N. Degradation of acetaldehyde and its precursors by *Pelobacter carbinolicus* and *P. acetylenicus*. PLoS One. 2014;**9**(9, 12): e115902. DOI: 10.1371/journal.pone.011590

[46] Bertsch J, Siemund AL, Kremp F, Muller V. A novel route for ethanol oxidation in the acetogenic bacterium *Acetobacterium woodii*: The acetaldehyde/ethanol dehydrogenase pathway. Environmental Microbiology. 2016;**18**:2913-2922. DOI: 10.1111/1462-2920.13082

[47] Weghoff MC, Bertsch J, Muller V. A novel mode of lactate metabolism in strictly anaerobic bacteria. Environmental Microbiology. 2015;**17**: 670-6777. DOI: 10.1111/1462-2920.12493

[48] Thauer RK. Biochemistry of methanogenesis: A tribute to Marjory Stephenson. Microbiology. 1998;**144**: 2377-2406. DOI: 10.1099/00221287-144-9-2377

[49] Thauer RK, Kaster AK, Seedorf H, Buckel W, Hedderich R. Methanogenic Archaea: Ecologically relevant differences in energy conservation. Nature Reviews Microbiology. 2008;**6**: 579-591. DOI: 10.1038/nrmicro1931

Chapter 4

Review of Mathematical Models for the Anaerobic Digestion Process

Borja Velázquez-Martí, Orlando W. Meneses-Quelal, Juan Gaibor-Chavez and Zulay Niño-Ruiz

Abstract

To describe anaerobic fermentation, many mathematical models have been suggested. A commonly accepted hypothesis in microbial growth is the speed of cellular reproduction, which is proportional to the concentration of cells at that instant. The constant of proportionality between the speed of growth and cell concentration is called cell growth rate, μ. In many occasions, the cell growth rate is considered constant. This leads to conclude that the concentration of cells versus time presents an exponential function. The consideration of this equation provides a good adjustment in the beginning of central phase of the anaerobic fermentation process. However, it moves away from the measurements when there is a limited reproduction due to lack of nutrients and competition between the cells in the environment. This produces a sigmoidal variation in concentration. To find a suitable fit function for all phases of the process, Gompertz proposes a model that considers the cell growth rate as variable. In this chapter, the Gompertz model, kinetic models, transference, and cone models are evaluated. Different adaptations to fit the variables to the obtained values in the experiments have been reviewed.

Keywords: mathematical model, Gompertz, fermentation, kinetic model, methane

1. Introduction

Anaerobic digestion is a biological process in which the organic matter in the absence of oxygen, and through the action of a group of specific bacteria, is broken down into a set of gaseous products, called biogas, formed by CH_4, CO_2, H_2, H_2S, etc. and in a digestate, which is a mixture of mineral substances (N, P, K, Ca, etc.) and compounds of difficult degradation [1]. One of the objectives of anaerobic digestion is the production of methane, which can be used as fuel. Anaerobic digestion is considered one of the most important and advantageous processes in the treatment of livestock manure and sludge residues. It represents a possibility to reduce its environmental impact while at the same time, providing a biofuel for local energy needs [2]. This process has been known for hundreds of years; however, it is still the object of research due to the great variability of the conditions in which it can be produced, diversity of raw materials, and influential factors.

Table 1 shows some of the most recent researches. In recent years, there has been an increasing interest in new raw fermentation materials, mainly

Author	Material	Pretreatment	Methane potential m^3 $kg^{-1}SV$
Bayrakdar et al. [4]	Chicken manure		0.272
Franco et al. [5]	Wheat straw + inoculum		0.229
Franco et al. [5]	Wheat straw + glucose + ac. Formic + inoculum*		0.276
Guo et al. [6]	Excessively withered corn straw + glucose		0.282
Li et al. [7]	Parton + sheep manure		0.152
Li et al. [7]	Paper + sheep manure		0.199
Mancini et al. [8]	Lignocellulose in general	N-methylmorpholine N-oxide	0.304
Martín Juárez et al. [9]	Microalgae + pig manure	Alkaline pretreatment with NAOH	0.377
Mustafa et al. [10]	Bagasse of sugarcane + inoculum*	Hydrothermal pretreatment	0.318
Vazifehkhoran et al. [11]	Wheat straw + sewage		0.314
Xu et al. [12]	Corn straw + *Bacillus Subtilis*	Microaerobic mesolithic	0.270
Zahan et al. [13]	*Gallinaza* (sawdust, wood shavings, and rice or straw husk) with yogurt serum		0.670
Aboudi et al. [14]	Dry sediment of sugar beet tails + pig manure		0.260
Dennehy et al. [15]	Food waste and pig manure		0.521
Glanpracha and Annachhatre [16]	Cassava pulp with pig manure		0.380
Marin Batista et al. [17]	Vinasse and chicken manure (chicken dung)		0.650
Aboudi et al. [18]	Dry beet granules of sugar beet + cow dung		0.280
Belle et al. [19]	Fodder radish with cow dung		0.200
Cestonaro et al. [20]	Sheep litter (mixture of rice husk with feces and urine) + cattle manure		0.171
Di Maria et al. [21]	Sludge from wastewater with fruit and vegetable waste		0.216
Fu et al. [22]	Corn straw + inoculum *	Thermophilic microaerobic	0.326
Fu et al. [23]	Corn straw + inoculum *	Secondary thermophilic microaerobic	0.381
Agyeman and Tao [24]	Food waste + livestock manure		0.467

**Inoculum is material obtained from the effluent of a previous biogas plant that ferments raw materials, such as manure from pigs, cows, sheep, chickens, and other animals, at mesophilic ranges.*

Table 1.
Values obtained from methane potential in various co-digestion processes.

lignocellulosic materials from agriculture, or waste such as paper and cardboard. So, co-digestion processes are being analyzed, which consist of improving methane production by mixing materials that ferment better together than separated due to the enriched microbial load; in this way, their nutritional needs are better complemented.

New inocula, such as the rumen, and its interaction with the raw material are also being examined, together with nutritional requirements. Pretreatment studies are being carried out along with thermal sequences in the processes, alternating thermophilic and mesophilic stages and evaluating the productivity, kinetics, and net energy balance. The microbiological identification involved in the fermentation according to the substrate and the followed thermal process also acquire interest.

One of the most discussed aspects is mathematical modeling. The objective of the modeling is to be able to establish characteristic parameters of the raw material and process conditions to predict the system's evolution over time, the performance obtained, and fermentation speed. In this study the most important models are evaluated.

Anaerobic digestion comprises a decomposition mechanism of organic matter based on three stages [3]: first a hydrolytic phase, in which polymers of long carbon chains are broken obtaining shorter acid chains, subsequently, an acetogenic phase, in which the short-chain acids obtained in the previous phase are transformed into acetic acid, and finally, a methanogenic phase, in which the acetic acid is transformed into methane.

Each of these stages is provided by a differentiated microbiological group. Each group takes as a substrate to the product generated in the previous phase. When the evolution of a microbial group is analyzed in a batch-type reactor, in batches, the variation of cell concentration varies, as shown in **Figure 1**.

Initially, the concentration of microorganisms responsible of digestion is small and evolves very slowly in this stage because it needs time to adapt. This phase is called *lag phase*, or lethargy. Subsequently, there is a very rapid increase in cell concentration called the growth phase. The growth phase ends when cell compete for substrate, causing a number of cell replications to equal deaths, so the number of living cells is stabilized. This phase is called the *stationary phase*. The stationary

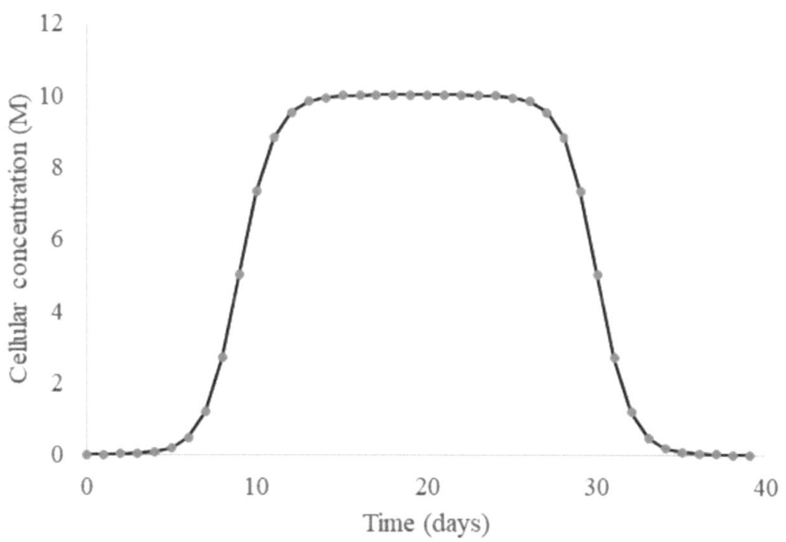

Figure 1.
Variation of cell concentration over time in a batch reactor.

phase ends when this battle for substrate causes a higher number of deaths than the number of reproductions, resulting in cell concentration to fall sharply. This phase is called the cell *death phase*.

From the practical point of view, it is only interesting to analyze the period between the beginnings of the fermentation to the stationary phase, appearing a curve similar to the sigmoid one. However, the sigmoid equation does not correctly fit the experimental results obtained.

2. Exponential model

A model widely used to describe the variation of cell concentration in the growth phase has been the exponential model. This model is based on the hypothesis that the speed of growth in an instant is proportional to the concentration of cells existing at that moment. This is expressed mathematically by Eq. (1), where X is the concentration of cells and μ is the constant of proportionality called cell growth rate:

$$\frac{dX}{dt} = \mu \cdot X \tag{1}$$

The development of Eq. (1) shows that, in the growth phase, the variation of cells follows an exponential curve:

$$\frac{dX}{X} = \mu \cdot dt$$

$$\int_{X_1}^{X_2} \frac{dX}{X} = \int_{t_{lag}}^{t} \mu \cdot dt$$

$$\ln \frac{X_2}{X_1} = \mu \cdot \left(t - t_{lag}\right)$$

$$X_2 = X_{\cdot 1} e^{\mu \cdot \left(t - t_{lag}\right)}$$

t_{lag} is the lag time. The cell growth rate has as unit the inverse of time (d^{-1}) and can be calculated experimentally with Eq. (2):

$$\mu = \frac{X_2 - X_1}{X_1 \cdot \left(t - t_{lag}\right)} \tag{2}$$

This model is not completely satisfactory because it has been verified that μ is not constant and it varies as time goes by. As competition for the substrate increases, the curve in **Figure 1** moves away from the exponential. To achieve a better fit, Monod proposed a model for calculating the cell growth rate as a function of the substrate concentration according to Eq. (3), where S is the substrate concentration at a given time, μ_{max} is the maximum rate of cell growth, and Ks is a constant called saturation:

$$\mu = \frac{\mu_{max} \cdot S}{K_s + S} \tag{3}$$

The Monod model proposes the existence of a maximum cell growth rate and a saturation constant that are characteristics of microbial species growing under defined conditions. The maximum growth rate is the one that occurs initially in the growth phase exponentially. When the substrate begins to be scarce, the rate decreases with respect to the maximum.

Along with the Monod model, there are others with the same style that can be observed in **Table 2**. In all of them, it can be seen that the maximum rate value considered in the exponential phase is minorized when the substrate concentration is low.

The relationship between the variations of cell concentration is always proportional to substrate consumption. The proportionality constant is called the biomass/substrate yield $Y_{x/s}$ and is defined by Eq. (4), where S_0 and S_1 are the initial and final substrate concentrations and X_0 and X_1 are the initial and final cell concentrations:

$$Y_{x/s} = \frac{X_1 - X_0}{S_0 - S_1} \tag{4}$$

If the initial concentration of substrate (So) is known, the variation of cell mass during the process is obtained from the biomass/substrate ratio of the process Yx/s. Limiting the decrease in the growth rate to a certain percentage of its maximum value allows calculating the time retention (TR) in a bioreactor batch.

$$z \cdot \mu_{\max} = \frac{\mu_{\max} S_1}{K_s + S_1} (0<z<1) \rightarrow S_1 = \frac{z}{1-z} \cdot K_s$$

$$Y_{x/s} = \frac{X_1 - X_0}{S_0 - S_1} \rightarrow X_1 = X_0 + Y_{x/s} \cdot (S_0 - S_1)$$

$$\ln \frac{X_1}{X_o} = \mu_{\max} \cdot \left(TR - t_{lag}\right) \rightarrow TR = t_{lag} + \frac{1}{\mu_{\max}} \ln \frac{X_1}{X_o}$$

The amount of product generated per unit volume and time (P) and methane in this case (M) are proportional to the variation of cell concentration (X). The proportionality constant $Y_{p/x}$ is called product/biomass yield:

$$Y_{p/x} = \frac{P_1 - P_0}{X_1 - X_0}$$

Type of model	Author	Model
Kinetic models without inhibition	Tessier	$\mu = \mu_{\max} \cdot \left(1 - e^{-S/K_s}\right)$
	Moser	$\mu = \mu_{\max} \frac{S^n}{K_s \cdot a + S^n}$
	Contois	$\mu = \mu_{\max} \frac{S}{BX + S}$
Kinetic models with inhibition	Andrews and Noak	$\mu = \mu_{\max} \frac{1}{K_s + S + \frac{S^2}{K_{is}}}$
	Webb	$\mu = \mu_{\max} \frac{S \cdot \left(1 + \frac{\beta \cdot S}{K_{is}}\right)}{K_s + S + \frac{S^2}{K_{is}}}$
	Aiba et al.	$\mu = \mu_{\max} \frac{S}{K_s + S} e^{-S/K_{si}}$
	Teissier	$\mu = \mu_{\max} \left[e^{-S/K_{si}} - e^{-S/K_s}\right]$
	Tseng and Wymann	$\mu = \mu_{\max} \frac{S}{K_s + S} - K_{si}(s - s_c)$

Table 2.
Variation models of the cell growth rate [25].

$$\frac{dM}{dt} = Y_{p/x} \cdot \frac{dX}{dt}$$

Since the variation of cell concentration is proportional to the concentration of cells at a given time, we have to.

$$\frac{dM}{dt} = Y_{p/s} \cdot \mu X$$

By developing the variation of cell concentration over time, it has been demonstrated that the amount of product obtained (methane) follows an exponential growth during the exponential growth of microorganisms. That is the reason because working in this phase with batch-type bioreactors is preferred for optimum performance. To do this, you must adjust the retention time to the duration of this stage.

X_0 represents the initial cell concentration in the reactor; X represents cell concentration at a time t, and t_{lag} is the time of lethargy or cellular adaptation:

$$\frac{dM}{dt} = Y_{p/s} \cdot \mu X_0 \cdot e^{\mu(t-tlag)}$$

$$M = Y_{p/s} \cdot X_0 \cdot \left(e^{\mu(t-tlag)} - 1\right)$$

whereas the value of $Y_{p/s} \cdot X_0$ is negligible compared to the exponential, that is $Y_{p/s} \cdot X_0 <<< Y_{p/s} \cdot X_0 \cdot e^{\mu(t-tlag)}$, the accumulated volume obtained in each experiment can be graphically represented with the model of Eq. (1), calculating the cell growth rate, the productivity of the substrate, and the optimum retention time for a greater use of energy:

$$M = Y_{p/s} \cdot X_0 \cdot e^{\mu(t-tlag)}$$

3. Model of Gompertz

Despite the practicality of the exponential model when complemented by the Monod equation, it is not completely satisfactory because it does not describe well the variation of cell concentration as the substrate is being consumed and the stationary phase approaches. Knowing how cell growth behaves in this area is significantly relevant if you want to use high retention times.

To find an adequate adjustment function for all phases of the process, Winsor [26] proposes to use an equation developed by Gompertz [27] in human demography. This proposes a model that considers the variable cell growth rate, as shown in Eqs. (5) and (6), where a and c are constants:

$$\frac{dX}{dt} = c \cdot \ln(a/X) \cdot X \tag{5}$$

$$\mu = c \cdot \ln(a/X) \tag{6}$$

According to Eq. (6), Gompertz moves radically away from the Monod approach, since the cell growth rate has no maximum. If there was a maximum, the derivative of Eq. (6) would be canceled at some point, something that does not happen:

$$\lim_{X\to 0}\mu = \lim_{X\to 0} c\cdot \ln\left(a/X\right) = \infty$$

$$\lim_{X\to \infty}\mu = \lim_{X\to \infty} c\cdot \ln\left(a/X\right) = -\infty$$

$$\frac{d\mu}{dt} = c\frac{X}{a}\cdot\left(\frac{-a}{X^2}\right) = \frac{-c}{X}$$

To obtain the function of cell concentration in time according to Gompertz, we must solve Eq. (5), which is a differential equation of separable variables:

$$\frac{dX}{X\cdot\ln\left(a/X\right)} = c\cdot dt$$

$$\int_{X0}^{X}\frac{dX}{X\cdot\ln\left(a/X\right)} = \int_0^t c\cdot dt$$

$$-\left[\ln\left(\ln\frac{a}{X}\right) - \ln\left(\ln\frac{a}{X_0}\right)\right] = ct$$

$$\ln\left(\frac{\ln\frac{a}{X_0}}{\ln\frac{a}{X}}\right) = ct$$

$$\frac{\ln\frac{a}{X_0}}{e^{ct}} = \ln\frac{a}{X}$$

Since a and X_0 are constants, the following consideration can be made:

$$\ln\frac{a}{X_0} = B = e^b$$

$$e^{e^{-ct+b}} = \frac{a}{X_0}$$

Therefore, Eq. (7) is obtained, which describes the cellular concentration in the reactor for each instant. This equation is the true contribution of the Gompertz:

$$X = a\cdot e^{\left[-e^{-ct+b}\right]} \tag{7}$$

When analyzing the limits in zero and infinity, we observe that the initial concentration of cells is $X1$ and that a represents an asymptote corresponding to the maximum cell potential, which would occur in the steady state:

$$\lim_{t\to 0} X = a\cdot e^{-B} = a\cdot e^{\ln\frac{X_0}{a}} = X_0$$

$$\lim_{t\to\infty} X = a$$

3.1 Considerations to the Gompertz model

If we accept the Gompertz model, Zwietering et al. [28] suggest modifications providing physical meaning to these variables. The rate of growth can be redefined as Eq. (8):

$$\frac{dX}{dt} = a\cdot e^{\left[-e^{-ct+b}\right]}\cdot\left(-e^{-ct+b}\right)\cdot -c = a\cdot c\cdot e^{\left[-e^{-ct+b}\right]}\cdot e^{-ct+b}$$

$$\frac{dX}{dt} = a \cdot c \cdot e^{\left[-e^{-ct+b}\right]} \cdot e^{-ct+b} \tag{8}$$

The instant in which the maximum growth velocity t_m occurs would be calculated from the first derivative of the velocity equal to zero, which is the same as the second derivative of the Gompertz Eq. (7). This implies that at that point where the growth speed is at maximum, the Gompertz function has a turning point:

$$\frac{d^2X}{dt^2} = a \cdot c^2 \cdot e^{\left[-e^{-ct+b}\right]} \cdot \left(e^{-ct+b}\right)^2 - a \cdot c^2 \cdot e^{\left[-e^{-ct+b}\right]} \cdot \left(e^{-ct+b}\right)$$

$$\frac{d^2X}{dt^2} = a \cdot c^2 \cdot e^{\left[-e^{-ct+b}\right]} \cdot \left(e^{-ct+b}\right) \cdot \left[\left(e^{-ct+b}\right) - 1\right]$$

$$\frac{d^2X}{dt^2} = a \cdot c^2 \cdot e^{\left[-e^{-ct_m+b}\right]} \cdot \left(e^{-ct_m+b}\right) \cdot \left[\left(e^{-ct_m+b}\right) - 1\right] = 0$$

$$-ct_m + b = 0$$

$$t_m = \frac{b}{c}$$

The concentration of cells where the maximum reproduction speed occurs is calculated by entering the value of t_m in Eq. (7), and it is shown that the growth rate where the reproduction speed is at maximum equals c:

$$X = a \cdot e^{\left[-e^{-ct_m+b}\right]} = a \cdot e^{\left[-e^{-c\frac{b}{c}+b}\right]} = \frac{a}{e}$$

$$\mu_m = c \cdot \ln\left(a/(a/e)\right) = c$$

The maximum reproduction speed value is obtained by substituting t_m in Eq. (8):

$$v_{\max} = \frac{dX_{tm}}{dt} = a \cdot c \cdot e^{\left[-e^{-ctm+b}\right]} \cdot e^{-ct+b} = a \cdot c \cdot e^{\left[-e^{-c\frac{b}{c}+b}\right]} \cdot e^{-c\frac{b}{c}+b} = \frac{a \cdot c}{e}$$

According to the previous thing, the curve tangent X in the point of inflection t_m has the form:

$$X = \frac{a \cdot c}{e} t + k$$

$$\text{Given the } t = t_m = \frac{b}{c} \text{ y } X_{tm} = \frac{a}{e,} \text{ so :}$$

$$\frac{a}{e} = \frac{a \cdot c}{e} \cdot \frac{b}{c} + k \rightarrow k = \frac{a}{e} - \frac{a \cdot b}{e} = \frac{a}{e}(1 - b)$$

$$X = \frac{a \cdot c}{e} t + \frac{a}{e}(1 - b) = \frac{a}{e} \cdot (ct + (1 - b))$$

If we define the latency time, t_{lag}, as the time in which the tangent line at the curve inflection point (point that coincides with maximum velocity) cuts the axis of the abscissa, we have that the latency time is in $X = 0$:

$$0 = ct_{lag} + (1 - b)$$

$$t_{lag} = \frac{(b-1)}{c}$$

From this equation, b can also be expressed as.

$$b = c \cdot t_{lag} + 1$$

$$\text{And } v_{\max} = \frac{a \cdot c}{e}, \text{ the result}$$

$$b = \frac{v_{\max} \cdot e}{a} \cdot t_{lag} + 1$$

Obtaining the Gompertz equation is Eq. (9). This equation has become popularized as the *modified Gompertz equation*:

$$X = a \cdot e^{\left[-e^{\frac{v_{\max} \cdot e}{a} \cdot \left(t_{lag} - t\right) + 1}\right]} \tag{9}$$

This equation has been used in current research, such as Bah et al. [29], Capson-Tojo et al. [3], Bayrakdar et al. [4], Mancini et al. [8], Martín Juárez et al. [9], and Li et al. [7].

To experimentally obtain the maximum reproduction speed and the latency time, X is measured as well as the reactor time. Next by defining the value of a as the maximum cell concentration obtainable, Eq. (9) then can be linearized:

$$\ln\left(\ln \frac{X}{a}\right) = -\frac{v_{\max} \cdot e}{a} t + \left(1 + \frac{v_{\max} \cdot e}{a} t_{lag}\right)$$

The latency time and the maximum speed of cellular reproduction will be characteristics of the microbial group in certain conditions.

3.2 Cumulative production curve of methane applying Gompertz

If we consider the product/biomass yield, we have.

$$Y_{p/x} = \frac{P_1 - P_0}{X_1 - X_0} = \frac{dM}{dX}$$

$$\frac{dM}{dt} = Y_{p/x} \frac{dX}{dt} \tag{10}$$

$$\frac{dM}{dt} = Y_{p/x} \cdot a \cdot c \cdot e^{\left[-e^{-ct+b}\right]} \cdot e^{-ct+b}$$

$$\frac{dM}{dt} = Y_{p/x} \cdot a \cdot c \cdot e^{\left[-e^{-\frac{v_{\max} \cdot e}{a} t + \frac{v_{\max} \cdot e}{a} \cdot t_{lag} + 1}\right]} \cdot e^{-\frac{v_{\max} \cdot e}{a} t + \frac{v_{\max} \cdot e}{a} \cdot t_{lag} + 1}$$

$$\frac{dM}{dt} = Y_{p/x} \cdot a \cdot c \cdot e^{\left[-e^{\frac{v_{\max} \cdot e}{a}\left(t_{lag} - t\right) + 1}\right]} \cdot e^{\frac{v_{\max} \cdot e}{a}\left(t_{lag} - t\right) + 1}$$

$$M = \int_0^t Y_{p/x} \cdot a \cdot c \cdot e^{\left[-e^{\frac{v_{\max} \cdot e}{a}\left(t_{lag} - t\right) + 1}\right]} \cdot e^{\frac{v_{\max} \cdot e}{a}\left(t_{lag} - t\right) + 1} dt$$

From Eq. (10), we obtain the cumulative methane production Eq. (11):

$$M = Y_{p/x} \cdot a \cdot e^{\left[-e^{\frac{v_{\max} \cdot e}{a}\left(t_{lag} - t\right)+1}\right]} \tag{11}$$

Taking limit when the time tends to infinity, it is shown that the methane potential produced is $Y_{p/x} \cdot a$:

$$\lim_{t\to 0} M = Y_{p/x} \cdot a \cdot e^{-B} = Y_{p/x} \cdot a \cdot e^{\ln \frac{X_1}{a}} = Y_{p/x} \cdot X_0$$

$$\lim_{t\to\infty} M = Y_{p/x} \cdot a$$

If we calculate the second derivative of the methane production curve and we equate to zero, then a maximum methane speed production point occurs:

$$\frac{d^2M}{dt^2} = 0$$

$$Y_{p/x} \cdot a \cdot c \cdot e^{\left[-e^{\frac{v_{\max} \cdot e}{a}\left(t_{lag} - t\right)+1}\right]} \cdot \left(-\frac{v_{\max} \cdot e}{a}\right) \cdot e^{\frac{v_{\max} \cdot e}{a}\left(t_{lag} - t\right)+1} \cdot \left(\left(-e^{\frac{v_{\max} \cdot e}{a}\left(t_{lag} - t\right)+1}\right) + 1\right) = 0$$

$$\frac{v_{\max} \cdot e}{a}\left(t_{lag} - t\right) + 1 = 0$$

$$t = \frac{a}{v_{\max} \cdot e} + t_{lag} = \frac{b}{c}$$

The maximum methane production rate is $v_{CH4\max}$:

$$v_{M\max} = Y_{p/x} \frac{a \cdot c}{e}$$

Lay et al. [30] proposed to modify the Gompertz Eq. (9) by applying the potential of producible methane, $M_e = Y_{p/x} \cdot a$, expressed as Eq. (12):

$$M = M_e \cdot e^{\left[-e^{\frac{v_{M\max} \cdot e}{M_e}\left(t_{lag} - t\right)+1}\right]} \tag{12}$$

Table 1 shows the values obtained from the methane potential in various co-digestion studies. All of them were carried out in mesophilic conditions, between 30 and 37°C. It can be observed that the production of methane in most cases ranges between 0.15 and 0.65 m^3 $kg^{-1}SV$. Based on this calculation, we could classify the digestion processes into three groups: (a) low-production processes, the amount of methane produced is between 0.15 and 0.30 m^3 $kg^{-1}SV$, (b) medium-production processes, the amount of methane produced is between 0.300 and 0.45 m^3 $kg^{-1}SV$, and (c) high-production processes, the amount of methane produced is greater than 0.45 m^3 $kg^{-1}SV$.

These types of productions and their energy equivalence mean that anaerobic digestion processes are considered more as a waste management and treatment process with a complementary energy product than as an alternative energy source to the problems derived from the limitation of fossil fuels.

3.3 Conclusions of the Gompertz model

The Gompertz model provides an equation that describes cell concentration over time in a fermentation process.

DOI: http://dx.doi.org/10.5772/intechopen.80815

To define this equation, it is necessary to obtain the value of three constants: *a* is the maximum cellular concentration, *b* is a constant that depends on the initial concentration of cells and *a*, and *c* is the value of the cell growth rate where the growth velocity is at maximum, that is, at the inflection point of the curve.

The Gompertz model implies that there is no maximum cell growth rate.

4. Kinetic models

The complexity of the Gompertz model and the problems that exist when applying the derivatives of the Monod and Contois equation have led some researchers to suggest models that do not focus on the growth rate but on the kinetics of substrate degradation or product formation. Brulé et al. [31] classify the kinetic models into four groups:

a. Reaction in a single step with first-order kinetics.

b. Two-step reaction with first-order kinetics.

c. Reaction in two speeds of a single step with first-order kinetics.

d. Reaction in two speeds of two steps with first-order kinetics.

4.1 One-step reaction with first-order kinetics

This model shows reaction rate is proportional to the amount of reagent, in this case substrate. So

$$\frac{dS}{dt} = k \cdot S \rightarrow S = S_0 \cdot e^{-k \cdot t}$$

where S is the amount of substrate at a time t, S_0 is the initial substrate amount, and k is the kinetic constant.

As the mass in the reaction is conserved, the mass of product M (methane) is calculated as

$$M = S_0 \cdot \left(1 - e^{-k \cdot t}\right)$$

Angelidaki et al. [32] used this kinetic type, relating the concentration of methane that is generated in a reactor with the maximum potential through the following equation:

$$\ln\left(\frac{M_e - M}{M_e}\right) = -k \cdot t$$

$$M = M_e \cdot \left(1 - e^{-k \cdot t}\right)$$

where M is the methane produced at a given time t, M_e is the value of the final methane production, and k is the constant of the hydrolysis rate.

Díaz et al. [33] evaluated the digestion of cellulose with manure by comparing the first-order equation, including in the equation the latency time (13) and the modified Gompertz equation. They concluded that both models did not offer significant differences in the coefficient of determination obtained in the models (r^2), neither in the methane potential predicted *Me* nor between the constant kinetics *k* and $v_{M\max}$. However, it shows that the first-order kinetic model provides a longer latency time. The maximum methane potential *Me* was between 0.30 and 0.33 m^3/kg SV:

$$M = M_e \cdot \left(1 - e^{-k\cdot\left(t - t_{lag}\right)}\right) \tag{13}$$

Zhang et al. [34] also compared the modified Gompertz equation and thefirst-order kinetic model according to Eq. (13). Zhang confirms that the first-order kinetic model provides longer latency times and methane potentials than Gompertz. However, it provides slightly lower coefficients of determination.

4.2 Two-step reaction with first-order kinetics

Shin and Song [35] considered anaerobic digestion as a two-step process that could work at different speeds. Although this comprises a complex hydrolytic, acetogenic, and methanogenic process, a more suitable kinetic model than the previous one would consist in first considering the formation of volatile fatty acids (*VFAs*) from the substrate *Se* and, subsequently, the conversion of these acids into methane (*M*).

The formation of volatile fatty acids depends on the substrate concentration, following first-order kinetics, where k_1 is the kinetic constant of transformation of the substrate to *VFA*, *S* is the substrate concentration, and S_{VFA} is the concentration of acid grades:

$$\frac{dS_{VFA}}{dt} = k_1 \cdot S$$

Given the $S = S_0 \cdot e^{-k_1 \cdot t}$, you have the equation:

$$\frac{dS_{VFA}}{dt} = k_1 \cdot S_0 \cdot e^{-k_1 \cdot t}$$

On the other hand, the elimination of the fatty acids will depend on the concentration of the same, also following first-order kinetics, being k_2 as the kinetic constant of transformation of the *VFA* to *M*.

According to the mass balance in the formation of the *VFA*, a differential equation of constant coefficients of first order (14) is obtained:

$$\begin{aligned} \frac{dS_{VFA}}{dt} &= k_1 \cdot S_0 \cdot e^{-k_1 \cdot t} - k_2 \cdot S_{VFA} \\ \frac{dS_{VFA}}{dt} + k_2 \cdot S_{VFA} &= k_1 \cdot S_0 \cdot e^{-k_1 \cdot t} \end{aligned} \tag{14}$$

such as

$$y' + a(x) \cdot y = b(x)$$

$$y = e^{-\int a(x)dx} \cdot \int b(x) \cdot e^{\int a(x)dx} dx + C \cdot e^{-\int a(x)dx}$$

The solution to Eq. (14) results

$$S_{VFA} = k_1 \cdot S_0 \cdot \frac{e^{-k_2 \cdot t} - e^{-k_1 \cdot t}}{k_2 - k_1}$$

From this equation, the accumulated methane production is obtained as

$$\frac{dM}{dt} = k_2 \cdot S_{VFA}$$

$$\frac{dM}{dt} = k_2 \cdot k_1 \cdot S_0 \cdot \frac{e^{-k_2 \cdot t} - e^{-k_1 \cdot t}}{k_2 - k_1}$$

$$M = S_0 \cdot \left(1 - \frac{k_1 e^{-k_2 \cdot t} - k_2 e^{-k_1 \cdot t}}{k_1 - k_2}\right)$$

4.3 Reaction in two speeds of a single step with first-order kinetics

The chemical composition of the substrates is generally heterogeneous and can be constituted by several fractions with different hydrolysis rates. This implies that we can consider the process as two parallel but independent mechanisms that occur simultaneously. If we define α as the relation between the amount of rapidly degradable substrate and the total a, k_F as the first-order kinetic constant for degradation of rapidly degradable substrate, and k_L as the first-order kinetic constant for the degradation of slowly degradable substrate, the amount of methane produced can be defined with the model used by Kusch et al. [36] or Luna del Risco [37]:

$$M = S_e \cdot \left(1 - \alpha \cdot e^{-k_F \cdot t} - (1 - \alpha) \cdot e^{-k_L \cdot t}\right)$$

Dennehy et al. [15] compared three different kinetic models to determine the most suitable to describe the kinetics of the discontinuous co-digestion of food waste and pig manure at 37°C: (1) first order, (2) Gompertz, and (3) two-speed one-step reaction with first-order kinetics. They showed that the three models provide similar determination coefficients; however, the RMSE (root of the mean of the squares of the errors) is significantly reduced when the two-speed digestion is considered. The worst RMSE was for the Gompertz model. The first-order kinetic model reduced the RMSE by 39%, and the first-order kinetic model with two speeds reduced the RMSE by 80%. The highest methane yields they obtained were 0.521 ± 29 m^3 CH4 kg^{-1} VS.

4.4 Reaction in two speeds of two steps with first-order kinetics

If we consider two steps in each of the fractions of which the substrate is composed, both for the rapidly degradable substrate fraction and for the slowly degradable substrate fraction, we can obtain the following equation:

$$M = S_e \cdot \left[\alpha \cdot \left(1 - \frac{k_{HF} e^{-k_{MF} \cdot t} - k_{MF} e^{-k_{HF} \cdot t}}{k_{HF} - k_{MF}}\right) + (1 - \alpha) \cdot \left(1 - \frac{k_{HL} e^{-k_{ML} \cdot t} - k_{ML} e^{-k_{HL} \cdot t}}{k_{HL} - k_{ML}}\right)\right]$$

Brulé et al. [31] evaluated the four kinetic models described, concluding that the models that consider an easy speed in both a step and two steps yield a reasonable estimate. In contrast, the model that considers two speeds with a single step produces overestimates. Therefore, it is considered inadequate. This overestimation is corrected by applying the two-step model at two speeds but complicates its application.

5. Model based on the transfer function

Several studies, such as Ghufran and Charles [38], Li et al. [39], or Zahan et al. [13], have used a function derived from the first-order kinetic model but which substitutes the kinetic constant for the ratio between the maximum and the methane velocity:

$$M = M_e \cdot \left(1 - e^{-k\cdot\left(t - t_{lag}\right)}\right)$$

$$M = M_e \cdot \left(1 - e^{-\frac{v_{\text{maxM}}}{M_e}\cdot\left(t - t_{lag}\right)}\right)$$

6. Cone model

On the other hand, researchers, such as Pitt et al. [40], El-Mashad [41], Li et al. [39], and Zahan et al. [13], analyzed the cone model. This model describes the fermentation according to Eq. (15):

$$M = \frac{M_e}{1 + (k \cdot t)^{-n}} \tag{15}$$

7. Comparison of models

For the evaluation of the models, most researchers usually use two statistics: (a) coefficient of determination of the fit (r^2) and (b) root of the mean of the squares of the errors (RMSE) calculated by Eq. (16), where M_{model} is the value of methane predicted by the model at an instant t and M_{ob} is the value of methane observed experimentally:

$$RMSE = \sqrt{\frac{\sum\left(M_{\text{model}} - M_{\text{ob}}\right)^2}{n}} \tag{16}$$

Pitt et al. [40], Ghufran and Charles [38], El-Mashad [41], Li et al. [39], and Zahan et al. [13] compared the modified Gompertz model, the first-order kinetic model, the transfer function model, and the cone model, for different types of substrates and combinations in co-digestion.

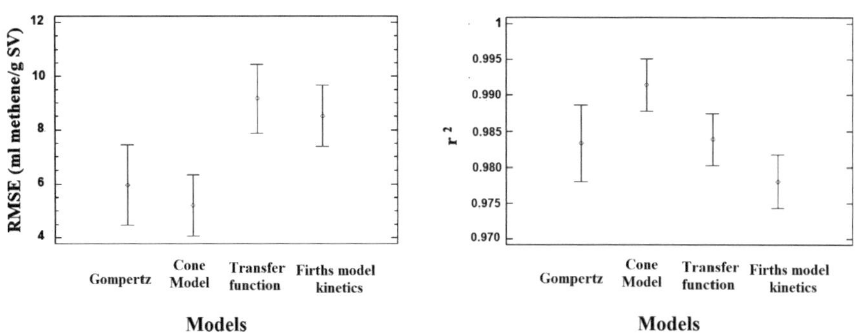

Figure 2.
LSD intervals of the analysis of variance at 95% confidence level for the comparison of the RMSE and the r^2 of the different models applied to the fermentation of different substances and combinations in co-digestion.

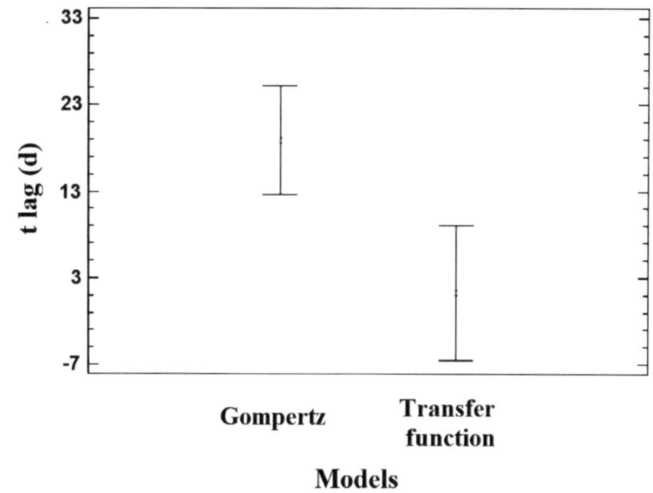

Figure 3.
LSD intervals of the analysis of variance at 95% confidence level for the comparison of the latency time of the different models applied to the fermentation of different substances and combinations in co-digestion.

Comparing the values of r^2, RMSE, and lag time provided by analysis of variance, the results shown in **Figures 2** and **3** were obtained.

As you can see, all the models provide high coefficients of determination, and there are few differences between them. The transfer model and the first-order kinetic model generally produce higher RMSE, so the modified Gompertz model and the cone model make more accurate estimates. However, the Gompertz model estimates higher latency periods.

8. Conclusion

In this research work, the most important kinetic models used to describe anaerobic fermentation have been developed. The comparison between them is a subject currently studied as demonstrated in recent publications. All of them provide high coefficients of determination; however, they present significant differences in the RMSE.

The production of methane in most cases ranges between 0.15 and 0.65 $m^3\ kg^{-1}SV$, under mesophilic conditions (30–37°C). However, digestion processes can be classified into three groups according to the methane production potential:

a. low-production processes, when the amount of methane produced is between 0.15 and 0.30 $m^3\ kg^{-1}SV$.

b. medium-production processes, when the amount of methane produced is between 0.30 and 0.45 $m^3\ kg^{-1}SV$.

c. high-production processes, when the amount of methane produced is greater than 0.45 $m^3\ kg^{-1}SV$.

The average lag time is 14 days.

The mean of the first-order kinetic constant is 0.11 d^{-1}.

Author details

Borja Velázquez-Martí[1*], Orlando W. Meneses-Quelal[1], Juan Gaibor-Chavez[2] and Zulay Niño-Ruiz[2]

1 Departamento de Ingeniería Rural y Agroalimentaria, Universitat Politècnica de Valencia, Valencia, Spain

2 Departamento de Investigación, Centro de Investigación del Ambiente, Universidad Estatal de Bolívar, Guaranda, Ecuador

*Address all correspondence to: borvemar@dmta.upv.es

References

[1] Deublein D, Steinhauser A. Biogas from Waste and Renewable Resources: An Introduction. 2nd ed. Weinheim, Germany: Wiley-VCH Verlag GmbH & Co. KGaA; 2010. 532 p. DOI: 10.1002/9783527632794

[2] Chynoweth DP, Wilkie AC, Owens JM. Anaerobic processing of piggery wastes: A review. In: ASAE Annual International Meeting; 12–16 July. Orlando, Florida, USA: St Joseph: American Society of Agricultural Engineers (ASAE); 1998. 38 pp

[3] Capson-Tojo G, Rouez M, Crest M, Trably E, Steyer J-P, Bernet N, et al. Kinetic study of dry anaerobic co-digestion of food waste and cardboard for methane production. Waste Management. 2017;**69**:470-479. DOI: 10.1016/j.wasman.2017.09.002

[4] Bayrakdar A, Sürmeli RÖ, Çalli B. Anaerobic digestion of chicken manure by a leach-bed process coupled with side-stream membrane ammonia separation. Bioresource Technology. 2018;**258**:41-47. DOI: 10.1016/j.biortech.2018.02.117

[5] Franco RT, Buffière P, Bayard R. Co-ensiling of cattle manure before biogas production: Effects of fermentation stimulants and inhibitors on biomass and methane preservation. Renewable Energy. 2018;**121**:315-323. DOI: 10.1016/j.renene.2018.01.035

[6] Guo J, Cui X, Sun H, Zhao Q, Wen X, Pang C, et al. Effect of glucose and cellulase addition on wet-storage of excessively wilted maize stover and biogas production. Bioresource Technology. 2018;**259**:198-206. DOI: 10.1016/j.biortech.2018.03.041

[7] Li W, Siddhu MAH, Amin FR, He Y, Zhang R, Liu G, et al. Methane production through anaerobic co-digestion of sheep dung and waste paper. Energy Conversion and Management. 2018;**156**:279-287. DOI: 10.1016/j.enconman.2017.08.002

[8] Mancini G, Papirio S, Lens PNL, Esposito G. Increased biogas production from wheat straw by chemical pretreatments. Renewable Energy. 2018;**119**:608-614. DOI: 10.1016/j.renene.2017.12.045

[9] Martín Juárez J, Riol Pastor E, Fernández Sevilla JM, Muñoz Torre R, García-Encina PA, Bolado Rodríguez S. Effect of pretreatments on biogas production from microalgae biomass grown in pig manure treatment plants. Bioresource Technology. 2018;**257**:30-38. DOI: 10.1016/j.biortech.2018.02.063

[10] Mustafa AM, Li H, Radwan AA, Sheng K, Chen X. Effect of hydrothermal and $Ca(OH)_2$ pretreatments on anaerobic digestion of sugarcane bagasse for biogas production. Bioresource Technology. 2018;**259**:54-60. DOI: 10.1016/j.biortech.2018.03.028

[11] Vazifehkhoran AH, Shin SG, Triolo JM. Use of tannery wastewater as an alternative substrate and a pre-treatment medium for biogas production. Bioresource Technology. 2018;**258**:64-69. DOI: 10.1016/j.biortech.2018.02.116

[12] Xu W, Fu S, Yang Z, Lu J, Guo R. Improved methane production from corn straw by microaerobic pretreatment with a pure bacteria system. Bioresource Technology. 2018;**259**:18-23. DOI: 10.1016/j.biortech.2018.02.046

[13] Zahan Z, Othman MZ, Muster TH. Anaerobic digestion/co-digestion kinetic potentials of different agro-industrial wastes: A comparative batch

study for C/N optimisation. Waste Management. 2018;**71**:663-674. DOI: 10.1016/j.wasman.2017.08.014

[14] Aboudi K, Álvarez-Gallego CJ, Romero-García LI. Evaluation of methane generation and process stability from anaerobic co-digestion of sugar beet by-product and cow manure. Journal of Bioscience and Bioengineering. 2016;**121**(5):566-572. DOI: 10.1016/j.jbiosc.2015.10.005

[15] Dennehy C, Lawlor PG, Croize T, Jiang Y, Morrison L, Gardiner GE, et al. Synergism and effect of high initial volatile fatty acid concentrations during food waste and pig manure anaerobic co-digestion. Waste Management. 2016; **56**:173-180. DOI: 10.1016/j.wasman.2016.06.032

[16] Glanpracha N, Annachhatre AP. Anaerobic co-digestion of cyanide containing cassava pulp with pig manure. Bioresource Technology. 2016; **214**:112-121. DOI: 10.1016/j.biortech.2016.04.079

[17] Marin Batista J, Salazar L, Castro L, Escalante-Hernández H. Co-digestión anaerobia de vinaza y gallinaza de jaula: alternativa para el manejo de residuos agrícolas colombianos. Revista Colombiana de Biotecnología. 2016; **18**(2):6-12. Retrieved from: https://dialnet.unirioja.es/descarga/articulo/5798936.pdf

[18] Aboudi K, Álvarez-Gallego CJ, Romero-García LI. Semi-continuous anaerobic co-digestion of sugar beet byproduct and pig manure: Effect of the organic loading rate (OLR) on process performance. Bioresource Technology. 2015;**194**:283-290. DOI: 10.1016/j.biortech.2015.07.031

[19] Belle AJ, Lansing S, Mulbry W, Weil RR. Anaerobic co-digestion of forage radish and dairy manure in complete mix digesters. Bioresource Technology. 2015;**178**:230-237. DOI: 10.1016/j.biortech.2014.09.036

[20] Cestonaro T, Costa MSS de M, Costa LA de M, Rozatti MAT, Pereira DC, Francisconi Lorin HE, et al. The anaerobic co-digestion of sheep bedding and ⩾50% cattle manure increases biogas production and improves biofertilizer quality. Waste Management. 2015;**46**:612-618. DOI: 10.1016/j.wasman.2015.08.040

[21] Di Maria F, Sordi A, Cirulli G, Micale C. Amount of energy recoverable from an existing sludge digester with the co-digestion with fruit and vegetable waste at reduced retention time. Applied Energy. 2015; **150**:9-14. DOI: 10.1016/j.apenergy.2015.01.146

[22] Fu SF, Wang F, Yuan XZ, Yang ZM, Luo SJ, Wang CS, et al. The thermophilic (55°C) microaerobic pretreatment of corn straw for anaerobic digestion. Bioresource Technology. 2015;**175**:203-208. DOI: 10.1016/j.biortech.2014.10.072

[23] Fu SF, Shi XS, Xu XH, Wang CS, Wang L, Dai M, et al. Secondary thermophilic microaerobic treatment in the anaerobic digestion of corn straw. Bioresource Technology. 2015;**186**: 321-324. DOI: 10.1016/j.biortech.2015.03.053

[24] Agyeman FO, Tao W. Anaerobic co-digestion of food waste and dairy manure: Effects of food waste particle size and organic loading rate. Journal of Environmental Management. 2014;**133**: 268-274. DOI: 10.1016/j.jenvman.2013.12.016

[25] Velázquez-Martí B. Aprovechamieno de la biomasa para uso energético. 1st ed. Barcelona, Spain: Reverté; 2017. 840 p. ISBN: 9788429147544

[26] Winsor CP. The Gompertz curve as a growth curve. Proceedings of the National Academy of Sciences. 1932;**18**: 1-8

[27] Gompertz B. On the nature of the function expressive of the law of human mortality, and on a new mode on determining the value of live contingencies. Philosophical Transactions of the Royal Society of London. 1825;**115**:513-585

[28] Zwietering MH, Jongenburger I, Rombouts FM, van't Riet K. Modeling of the bacterial growth curve. Applied and Environmental Microbiology. 1990; **56**(6):1875-1881

[29] Bah H, Zhang W, Wu S, Qi D, Kizito S, Dong R. Evaluation of batch anaerobic co-digestion of palm pressed fiber and cattle manure under mesophilic conditions. Waste Management. 2014;**34**(11):1984-1991. DOI: 10.1016/j.wasman.2014.07.015

[30] Lay JJ, Li YY, Noike T. Effect of moisture content and chemical nature on methane fermentation characteristics of municipal solid wastes. Doboku Gakkai Ronbunshu. 1996;**1996**(552): 101-108. DOI: 10.2208/jscej.1996. 552_101

[31] Brulé M, Oechsner H, Jungbluth T. Exponential model describing methane production kinetics in batch anaerobic digestion: A tool for evaluation of biochemical methane potential assays. Bioprocess and Biosystems Engineering. 2014;**37**(9):1759-1770. DOI: 10.1007/ s00449-014-1150-4

[32] Angelidaki I, Alves M, Bolzonella D, Borzacconi L, Campos JL, Guwy AJ, et al. Defining the biomethane potential (BMP) of solid organic wastes and energy crops. A proposed protocol for batch assays. Water Science and Technology. 2009;**59**(5):927-934. Retrieved from http.//wst.iwaponline. com/content/59/5/927.abstract

[33] Díaz I, Donoso-Bravo A, Fdz-Polanco M. Effect of microaerobic conditions on the degradation kinetics of cellulose. Bioresource Technology. 2011;**102**(21):10139-10142. DOI: 10.1016/j.biortech.2011.07.096

[34] Zhang H, Luo L, Li W, Wang X, Sun Y, Sun Y, et al. Optimization of mixing ratio of ammoniated rice straw and food waste co-digestion and impact o f trace element supplementation on biogas production. Journal of Material Cycles and Waste Management. 2018;**20**(2): 745-753. DOI: 10.1007/s10163-017-0634-0

[35] Shin HS, Song YC. A model for evaluation of anaerobic degradation characteristics of organic waste: Focusing on kinetics, rate-limiting step. Environmental Technology. 1995;**16**:775-784. DOI: 10.1080/ 09593331608616316

[36] Kusch S, Oechsner H, Jungbluth T. Biogas production with horse dung in solid-phase digestion systems. Bioresource Technology. 2008;**99**: 1280-1292. DOI: 10.1016/j. biortech.2007.02.008

[37] Luna del Risco M, Normak A, Orupold K. Biochemical methane potential of different organic wastes and energy crops from Estonia. Agronomy Research. 2011;**9**:331-342

[38] Ghufran R, Charles B. The use of a specific function to estimate maximum methane production in a batch-fed anaerobic reactor. Journal of Chemical Technology and Biotechnology. 2004; **79**(10):1174-1178. DOI: 10.1002/ jctb.1107

[39] Li K, Liu R, Sun C. Comparison of anaerobic digestion characteristics and kinetics of four livestock manures with different substrate concentrations. Bioresource Technology. 2015;**198**: 133-140. DOI: 10.1016/j.biortech. 2015.08.151

[40] Pitt RE, Cross TL, Pell AN, Schofield P, Doane PH. Use of in vitro gas production models in ruminal kinetics. Mathematical Biosciences.

1999;**159**(2):145-163. DOI: 10.1016/
S0025-5564(99)00020-6

[41] El-Mashad HM. Kinetics of methane production from the codigestion of switchgrass and Spirulina platensis algae. Bioresource Technology. 2013; **132**:305-312. DOI: 10.1016/j.biortech.2012.12.183

Section 3

Anaerobic Digestion Improvement and Evaluation

Chapter 5

Biomass Pretreatment for Enhancement of Biogas Production

Tamilarasan Karuppiah and Vimala Ebenezer Azariah

Abstract

Biomass is a renewable energy source developed from living or recently living plant and animal materials, which can be used as fuel. The main components present in biomass are polymers such as carbohydrate, protein, cellulose, lignin and fat. Biogas is produced when the biomass is anaerobically degraded by micro-organisms. The process of anaerobic digestion (AD) takes place in four steps: hydrolysis, acidogenesis, acetogenesis, and methanogenesis. The hydrolysis step is rate limiting due to the presence of complex polymers in biomass. Pretreatment is a process in which the biomass is made ready for microbial attack. This pretreatment can be physical operations such as communition, irradiation etc.; chemical treatment with alkali, acids, wet oxidation etc.; biological pretreatment, by fungi or enzymes; or a combination of these processes. During the pretreatment process, the compact structure of biomass is disrupted and exposed which.

Keywords: biomass, biogas production, anaerobic digestion, pretreatment, technologies

1. Introduction

Rapidly increasing energy demands worldwide has resulted in tremendous depletion of fossil fuel resources. This makes it necessary to find alternative energy sources which have a minimum impact on the environment. In this context, biogas is one of the sustainable energy sources that can be produced from many types of biomass including waste. AD technology is one of the most promising technologies, having the potential to convert various biomass into methane-rich biogas, a carbon-neutral alternative to fossil fuels. In addition, AD technology has a number of benefits including solids reduction, decreased odor, reduced greenhouse gas emissions, and increased income from non-market benefits compared to conventional waste treatment systems [1, 2]. In Germany, which is the leading country in this field, greater than 50% of the biogas potential results from energy crops treated in over 7000 biogas plants [3]. AD has wide application in sludge stabilization due to its low cost, energy recovery and minimized biosolids production.

AD system utilizes anaerobic microorganisms to convert the organic matter in the biomass, into biogas in an oxygen free environment. Biogas is the main byproduct of AD and contains about 60% methane by volume. Digestate is produced as a byproduct, which after an appropriate treatment can have agricultural applications as fertilizer [4]. It reduces organic matter to more stable solids by complex biochemical reactions. There are three consecutive steps of biological process in AD. The first step involves hydrolysis of complex organic matter into simpler compounds. The second step is the acidogenesis, which involves conversion of these organics to form organic acids and hydrogen. The final step is methane and carbon dioxide production from organic acids and hydrogen, by *methanogens.* The high methane content makes biogas a useful fuel that can displace natural gas in pipelines or be converted to electricity and heat. AD typically require long residence times, as certain anaerobic microorganisms have slow rate of growth. Long residence times lead to large volumes of tanks. Therefore, to improve digestion efficiency, the most efficient approach is to disrupt the chemical bonds in the material prone to hydrolysis [5]. Other factors limiting its performance are slow hydrolysis, low biodegradability, inhibition due to toxic compounds and toxic intermediates formed and poor methanogenesis. To overcome this recalcitrant property and to improve the degradation rate, a pretreatment prior to the AD process is introduced. Thus the goal of a pretreatment is to open up the structure of the substrate, making it more accessible for enzymatic attack [6] which aids in increasing biogas yield. The effects of various pretreatment methods highly differ, depending on the characteristics of the substrates and the pretreatment type. Recently, a lot of interest has been devoted to biomass disintegration and solubilization techniques in order to overcome the biological limitations of anaerobic digestion. The pretreatment techniques include mechanical treatment [7], ultrasonic treatment [8, 9] and biological hydrolysis with enzymes [10–12], alkaline treatment [13], oxidative treatments using ozone [14, 15], microwave irradiation [5, 16, 17], thermal treatment [18] thermochemical [19], sono-thermal [20–22] etc.

2. Microbiology of anaerobic digestion

AD process is mediated through four main steps—hydrolysis, acidogenesis, acetogenesis and methanogenesis. These are carried out by a consortium of microorganisms: acidogenic bacteria, acetogenic bacteria and methanogenic bacteria [23]. The microbial community of the anaerobic process is very complex. There are two prokaryotic kingdoms that closely interact with each other: *Bacteria* and *Archaea.* The first step involves hydrolysis of complex organic matter into simpler compounds. In the second step, the acidogenesis of these organics take place to form organic acids and hydrogen. In the final step, methane and carbon dioxide are produced from organic acids and hydrogen by *archael* methanogens.

Figure 1 summarizes the overall process of AD. Organic matter consists of particulate, water-insoluble polymers such as carbohydrates, lipids and proteins. Insoluble polymers cannot penetrate cellular membranes and are therefore not directly available to the microorganisms. During hydrolysis, appropriate strains of hydrolytic bacteria excrete hydrolytic enzymes [23] which break up the insoluble polymers to soluble mono and oligomers. Carbohydrates are converted to sugars, lipids are broken down to long-chain fatty acids and proteins are split into amino acids [24]. These soluble molecules are converted by acidogens to acetic acid and other longer volatile fatty acids, alcohols, carbon dioxide and hydrogen on acidogenesis. The foremost acids produced are acetic acid (CH_3COOH), propionic acid (CH_3CH_2COOH), butyric acid ($CH_3CH_2CH_2COOH$), and ethanol (C_2H_5OH). Other acid formers are *Clostridium, Peptococcusanerobus, Lactobacillus*, and *Actinomyces*. The next process is acetogenesis during which, the longer volatile fatty acids and alcohols are oxidized by proton-reducing acetogens to acetic acid and hydrogen. An acetogenesis reaction is shown below:

$$C_6H_{12}O_6 \rightarrow 2C_2H_5OH + 2CO_2 \tag{1}$$

In the last step of the process, methanogens use acetic acid or carbon dioxide and hydrogen, to produce methane and carbon dioxide. For mesophilic bacteria, the optimal methane production rate is mostly reached at 35–37°C. The thermophilic methanogens differ from the mesophilic one and their maximum methanogenic activity is reached at about 55°C. A thermophilic digestion process can sustain a higher organic loading compared to a mesophilic one. But the thermophilic process produces gas with a lower methane concentration [25] and is more sensitive to toxicants [26]. Methanogens are also sensitive towards changes in temperature than the other species, because of their slower growth rate in the reactor environment. Methanogenesis occurs at neutral pH- in the range of 6.5–7.5, although optimum lies at pH 7.0–7.2 [26]. If, for example, a temperature shift affects the methanogens negatively, there can be a build-up of volatile fatty acids (VFAs). This lowers the pH which further affects the methanogens in a negative way which leads to a vicious circle of negative feedback. The methanogenesis reactions can be expressed as follows [27] in Eqs. (2)–(4):

$$CH_3COOH \rightarrow CH_4 + CO_2 \tag{2}$$

$$2C_2H_5OH + CO_2 \rightarrow CH_4 + 2CH_3COOH \tag{3}$$

$$CO_2 + 4H_2 \rightarrow CH_4 + 2H_2O \tag{4}$$

The digestion efficiency and its stability can vary significantly depending upon the wastewater characteristics and type and design of the treatment system. The longer a substrate is kept under proper reaction conditions, the more complete its degradation will become. Longer retention time demands the provision of reactor with large volume for a given amount of substrate to be treated. On the other hand,

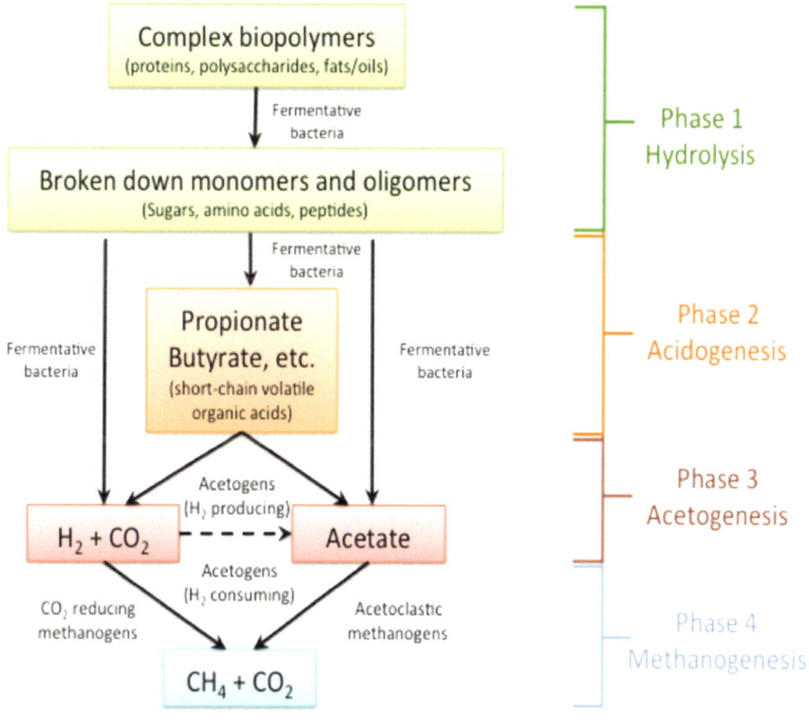

Figure 1.
Schematic representation of anaerobic digestion (source: https://www.e-education.psu.edu/egee439).

with shorter retention time washout of microorganism takes place with a lower overall degradation [25]. Therefore, these two effects have to be balanced in the design of AD for the efficient and proper operation of the full scale reactor. This needs operation of AD through skilled supervision for optimal performance.

3. Need for pretreatment

Several renewable matters have been tried for biogas production which are classified into crop biomass such as maize, wheat, barley, sweet sorghum, etc.; organic wastes such as municipal solid waste, municipal and industrial wastewater sludge, animal manure, and residues from various processing; energy crops like sunflower, rape, jatropha, etc.; crop residues which include banana stem, barley straw, rice straw, softwood spruce, etc.; and non-conventional biomass like glycerol, microalgae, etc. [28–34]. **Figures 2–4** show the effect of pretreatment of lignocellulosic, sludge and macroalgal biomass respectively.

The diverse composition of lignocellulose biomass and the interactions between fractions make its structure very complex and resistant to deconstruction. Cellulose and hemicellulose are polysaccharides that can be hydrolyzed to simple sugars. Lignin which acts as a support to the cell structure, embedding cellulose and hemicellulose, hinders the susceptibility to microbial attack during hydrolysis process [35]. The aim of pretreatment is to break the lignin layer that protects the cellulose and hemicellulose, in order to make the biomass more accessible for digestion [6]. Pretreatment also helps to decrease the crystallinity of cellulose and to increase the porosity. Furthermore, biomass such as fruit wastes is easily degraded but result in low yield due to the presence of inhibitors.

Keratin, which is present in horns and feathers, is an insoluble protein in which the polypeptides chain is tightly packed and highly cross-linked with disulfide bonds, hydrogen bonds, and hydrophobic interactions [36]. This insoluble protein is extremely resistive to the proteolytic enzyme action, which is a major hindrance in the biological processing of these wastes. For such biomass, if the crosslinking between the polypeptides chain breaks, the keratin becomes more accessible and easier to digest. Contrarily, while keratin-rich waste is pretreated using a strong

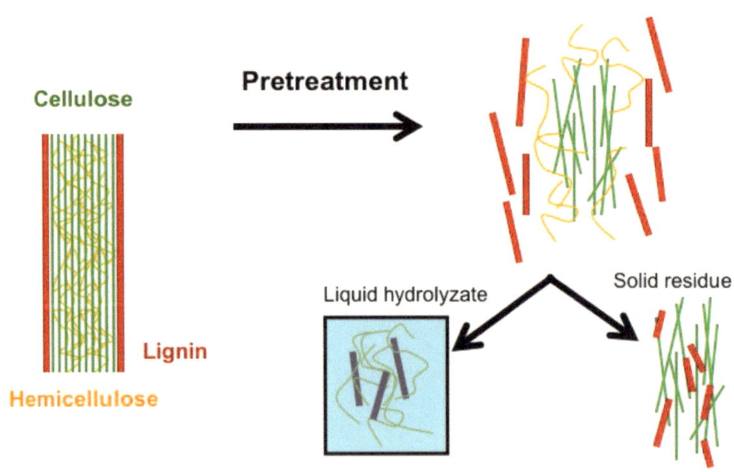

Figure 2.
Effect of pretreatment on lignocellulosic biomass (source: https://www.e-education.psu.edu/egee439/node/653).

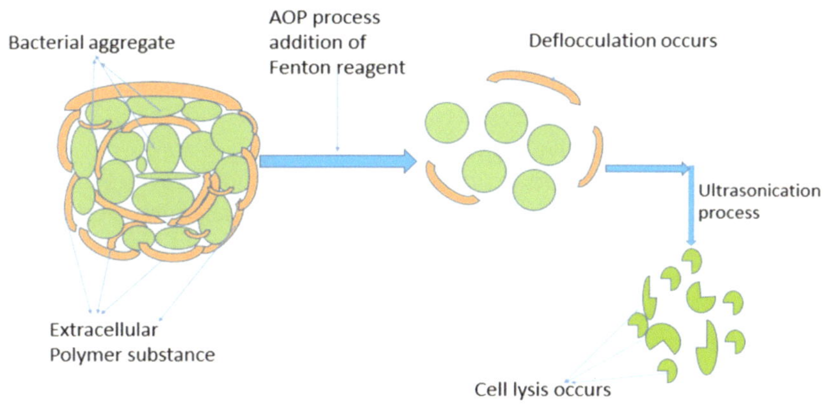

Figure 3.
Effect of pretreatment on sludge biomass.

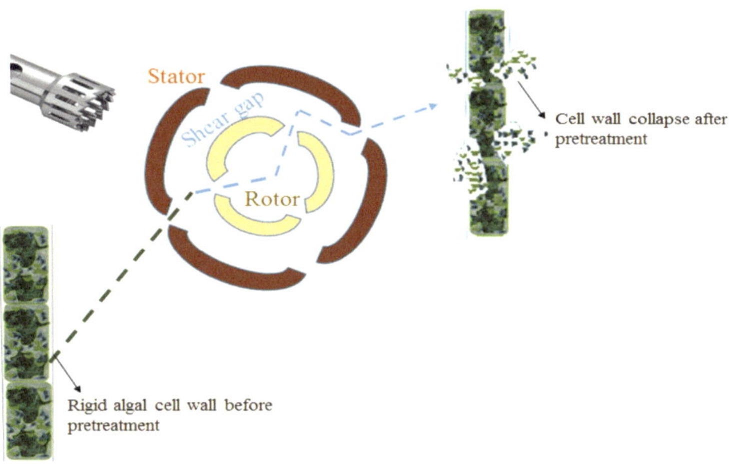

Figure 4.
Effect of pretreatment of macroalgal biomass.

acid, alkali, or other harsh physicochemical methods, severe degradation and destruction of the keratin occurs [37].

Activated sludge, a bio product of aerobic wastewater treatment, can be a better raw material for generating energy because of its high organic content [38]. Secondary wastewater sludge consists of numerous microbial cells, the cell walls of which act as barriers against exo-enzyme degradation. Besides microbial cells, exocellular polymeric substances (EPS) comprise a major organic fraction in activated sludge floc structure and binding mechanisms of EPS to cations appear to be a significant factor determining the digestibility of activated sludge. Hence hydrolysis becomes the rate-limiting step and degree of degradation achieved is limited to 30–35% chemical oxygen demand (COD) reduction in conventional anaerobic sludge treatment [23]. Pretreatment of sludge is required to rupture the cell wall and to facilitate the release of intracellular matter into the aqueous phase, which improves the biodegradability thereby enhancing the AD with lower retention time and with higher biogas production [20].

The macroalgal cell envelope made of thick and hard layer composed of complex proteins and carbohydrates with more mechanical power and high chemical resistance, restricts the attack of the biopolymers by methanogenic bacteria during

AD [39]. Pretreatment leads to improvement in the liquefaction process, enhancing the biopolymer release [28]. Several pretreatment methods have been reported in detail, aiming to make these biomass viable to digestion by microorganisms, and increase the biogas yield. It is necessary to carry out the pretreatment at mild conditions to prevent excessive sugar degradation.

Several pretreatment processes such as ball mill [40], microwave irradiation [2], sodium hydroxide [13], steam explosion [41], ultrasonic [42], biological [43], ozonation [14] have been shown to enhance biodegradability of biomass by promoting the hydrolysis process. Since most available articles are addressed based on pretreatment of lignocellulosic biomass, this chapter is mainly focused towards sludge pretreatment.

4. Pretreatment technologies

The lower hydrolysis rates during conventional AD process, results in higher hydraulic retention time (HRT) in the digester and larger digester volume, constitutes the prime drawbacks of the conventional AD [6]. The non-availability of the readily biodegradable, soluble organic matters and lower digestion rate constantly necessitates the pretreatment of sludge. Pretreatment of biomass enhances the AD, with lower retention time and with higher biogas production [17]. With the advancements in various pretreatment techniques like thermal, chemical, mechanical, biological and physical and several combinations such as physicochemical, biological–physicochemical, mechanical–chemical and thermal–chemical, biodegradability of sludge can be enhanced by several orders. Extensive research has been carried throughout the world to establish the best economically feasible pretreatment technology to enhance the digestibility of biomass [12].
Tables 1 and **2** show the specific energy consumed and methane yield with various chemo-mechanical and physico-chemical pretreatment.

4.1 Physical

In physical pretreatment, the structure of the biomass gets altered and the size of the particles reduced, by the application of physical force. This leads to an increase in the surface area of the particles thereby making it susceptible to microbial and enzymatic attacks, which enhances the AD process for methane production [61]. Physical pretreatment may be done by employing microwave irradiation, sonication, mechanical beating, deflaking, dispersing, extruding, refining, milling, and cavitation etc. [62].

4.1.1 Milling

Milling pretreatment is carried out, especially for lignocellulose and algal biomass to reduce the size of the substrate to break open the cellular structure, and improve their bio accessibility to the cell tissues, by increasing the specific surface area of the biomass [40]. Particle size reduction not only increases the rate of enzymatic degradation, but also reduces viscosity in digesters thus making mixing easier and can reduce the problems of floating layers. For effective hydrolysis of lignocellulose, beta particle size of 1–2 mm has been recommended [63]. Using three batch reactors, Motte et al. [40] demonstrated, treating straw particle milled to different sizes 0.25 mm, 1 mm and 10 mm followed during 62 days. They achieved the highest methane production for straw with 10 mm particle size (192 $\pm$ 25 Nm L/g VS) which was associated with a straw biodegradability of 43%.

S. No.	Name of the pretreatment	Specific energy consumed (KJ/kg TS)	Solubilization achieved (%)	Biomethane yield	References
1	Disperser + alkali	4544	24	1391 ml	Rani et al. [21]
2	Thermo chemo disperser	3360.94	18.6	0.455 L/g VS	Kavitha et al. [44]
3	Chemo disperser	5013	20	0.522 L/g VS	Poornima Devi et al. [45]
4	Sono alkaline	4172	59	0.108 ml/g VS removed	Rani et al. [46]
5	Thermo chemo sonic	5290.5	27	0.413 g COD/ g COD	Kavitha et al. [47]
6	Citric acid + ultrasonic	171.9	22.7	0.435 L/g VS	Gayathri et al. [29]
7	Fenton + ultrasonic	641	34.4	0.3 g COD/g COD	Kavitha et al. [48]
8	Thermo chemo sonic	5500	35	0.60 g COD/g COD	Kavitha et al. [49]
9	Disperser + microwave	18,000	22	0.28 g COD/g COD	Kavitha et al. [50]
10	Chemo mechanical	7377	38	50 ml/g VS removed	Kavitha et al. [51]
11	Sonic mediated biological	2.45	23	0.19 d1	Kavitha et al. [52]
12	Chemo thermo disperser	174	60	0.84 g COD/g COD	Kavitha et al. [43]
13	Surfactant sonic	5120	24.7	0.24 g/g COD	Ushani et al. [53]
14	Chemo disperser	3312.6	15	0.14 g COD/g COD	Tamilarasan et al. [28]
15	Surfactant + sonic	5400	26	0.6 g/g COD	Santhi et al. [54]
16	Disperser + bacterial	9.5	22.4	0. 279 g COD/ g COD	Banu et al. [55]
17	Ultrasound + microwave	16,700	33.2	0.3 L/g COD	Kavitha et al. [56]
18	Surfactant + sonic	9600	23.9	0.239 g/g COD	Tamilarasan et al. [57]

Table 1.
Specific energy consumed and methane yield with various chemo-mechanical pretreatment.

4.1.2 Cavitation

The most frequently applied cavitation techniques include acoustic cavitation, which is produced by passing ultrasonic waves through the liquid medium and hydrodynamic cavitation produced using hydraulic systems. In acoustic cavitation, microbubbles called cavitation were developed when the ultrasound waves propagate in a liquid medium, due to a repeating pattern of compressions and

S. No.	Name of the pretreatment	Specific energy consumed (KJ/kg TS)	Solubilization achieved (%)	Biomethane yield	References
1.	Microwave	1844	18.6	0.162 ml/g VS removed	Rani et al. [33]
2.	Microwave + citric acid	14,000	31	0.615 L/g VS	Ebenezer et al. [38]
3.	Microwave + surfactant	14,000	28	0.47 L/g VS	Ebenezer et al. [58]
4.	Microwave + H_2O_2	18,600	56	0.323 L/g VS	Eswari et al. [59]
5.	H_2O_2 + microwave	18,910	46.6	250 ml/g VS	Eswari et al. [60]
6.	Thermo ozone	141.02	30.4	0.32 g COD/g COD	Kannah et al. [1]

Table 2.
Specific energy consumed with various physico-chemical pretreatment.

rarefactions. These cavitation expand to unstable size, and then rapidly collapse resulting in temperatures up to 5000 K and pressures up to 180 MPa. The rapid collapse of a numerous microbubbles generates powerful shear forces in the surrounding liquid, which damages the cell walls of microorganisms [21, 53]. However, higher sonication power level is reported to adversely affect the pretreatment process. At higher power level, bubbles are formed near the tip of the ultrasound transducer, which hinders the transfer of energy to the liquid medium [64].

In the ultrasonic pretreatment study on waste activated sludge (WAS), Apul & Sanin [7] investigated an improvement in anaerobic biodegradability at 15 min of sonication. They achieved an increase in daily biogas production and methane production by 49 and 74% respectively compared to control in semi continuous reactors at a solid retention time (SRT) of 15 days and organic loading rate of 0.5 kg/m^3 d. Zeynali et al. [42] studied the efficiency of ultrasonic pretreatment in improving biogas production from fruits and vegetable waste. They adopted three sonication times of 9, 18, 27 min operating at 20 kHz and amplitude of 80 μm on the substrate. The highest methane yield they obtained was at 18 min sonication with specific energy 2380 kJ/kg TS (Total solids) for a 12 d batch period, while longer exposure to sonication led to lower methane yield. The energy content of the biogas obtained by them was twice that of input energy for sonication. Alzate et al. [65] reported that, the sonication applied to macro algae at a specific energy input of 75 MJ/kg TS produced just 20% of the methane production. Upon increasing the specific energy to about 100–200 MJ/kg TS, they reported an increase in the methane production rate between 80 and 90%.

In hydrodynamic systems, cavitation is generated by forcing fluid flow through cavitating devices, where pressure substantially drops. Many microbubbles formed as a consequence of this pressure drop subsequently collapse. The collapse of the cavitation, results in release of large magnitudes of energy which helps in dissolution of biomass and makes it more suitable for subsequent bacterial decomposition, improving biogas yield during the AD process [66]. They investigated the application of hydrodynamic cavitation (HC) for the pretreatment of wheat straw with an objective of enhancing the biogas production. They observed the methane yields of 31.8 ml with untreated wheat straw, 77.9 ml with HC pre-treated wheat straw and a maximum yield of 172.3 ml with the combined pre-treatment using KOH and HC.

4.1.3 Microwave irradiation

During microwave irradiation the destruction of the microbial cells is caused by the disruption of the chemical (hydrogen) bonds in the cell walls and membranes, by polarized parts of macromolecules aligning with the poles of the electromagnetic field, which results in denaturation. Microwaves can induce an athermal effect in addition to their thermal effect due to dipole orientation, which results in possible breakage of hydrogen bonds and subsequently leads to the disintegration of the floc matrix [17]. They observed that, microbial cells exposed to MW showed greater damage at similar applied temperatures compared to conventional heating. Rincón et al. [67] studied the effect of a MW pre-treatment on olive mill solid residue to enhance its anaerobic digestibility. They carried out the experiment at a power of 800 W and temperature 50°C and observed a maximum methane yield of 395 ± 1 ml CH_4/g VS for an applied specific energy 7660 kJ/kg TS. Beszédes et al. [16] focused on the effects of MW irradiation at different power levels on biodegradation and subsequent AD of sludge from the dairy and meat industry. Compared to their results obtained from conventional heat treatment of the same sludge, the MW treatment proved to increase the methane yield.

4.1.4 Extrusion

In extrusion pretreatment, the biomass is allowed to experience heat, compression and shear force, which creates physical damage and chemical alterations of biomass cells while passing through the *extruder*. The extruder arrangement consists of single or twin screws that spin into a tight barrel, which is equipped with temperature control. When a biomass material passes through the barrel, it is exposed to friction and vigorous shearing causing an increase in temperature and pressure. When it exits the finishing end, the biomass material experiences a pressure release, which causes structural changes in the processed biomass enabling easy digestion in the subsequent step [61].

Maroušek [68] evaluated extrusion parameters of pelleted hay for maximal cumulative biogas production, and reported that, at optimal conditions of pressure 1.3 MPa, reaction time 7 min, and 8% dry matter, the maximal biogas production was 405 m^3/ton TS (with 52.3% methane), which was about a 33% increase over the biogas yield of control. Novarino and Zanetti [69] employed extrusion pretreatment to improve biogas production from the organic fraction of municipal solid waste, resulting in a biogas yield of 800 L/kg VS containing about 60% methane content.

4.2 Thermal

Thermal pretreatment improves hydrolysis, with increased methane yield during subsequent anaerobic digestion. A wide range of temperatures has been studied, ranging from 60 to 270°C, but temperatures above 200°C have been found responsible for the production of recalcitrant soluble organics or toxic/inhibitory intermediates during the pretreatment process [70]. Many studies employed at an optimum thermal range of 160–180°C for hydrolysis of wastewater sludge have proved an increase in methane yield during AD. Higher temperatures lead to a sharp reduction in biodegradability of sludge hydrolysate, due to production of recalcitrant soluble organics or toxic/inhibitory intermediates during the process [71]. The effect of thermal treatment of anaerobic sludge on the disintegration of the remaining organic fraction was evaluated by Borges and Chernicharo [18]. At 75°C, they observed an increase of 30–35 times increase in the concentrations of protein, carbohydrate, lipid and COD and an increase of 50% in the biogas production, thus characterizing a higher biodegradability of the remaining organic fraction.

4.3 Chemical

4.3.1 Acid

Acid pretreatment causes sludge disintegration and cell lysis which releases the intracellular organics, which become more bioavailable and thus increases the rate and efficiency of the digestion process [17]. In lignocellulosic biomass, the pretreatment results in the disruption of the Van der Waals forces, hydrogen bonds and covalent bonds that hold together the biomass components, which consequently causes the breaking of hemicellulose and the reduction of cellulose [72]. Devlin et al. [73] showed the improved effects of HCl pretreatment at pH 2 on subsequent digestion of WAS. In semi-continuous digestion experiments conducted for 12 day hydraulic retention time at 35°C, they found a 14.3% increase in methane yield compared to untreated WAS. Taherdanak et al. [74] used dilute sulfuric acid pretreatment, to improve the biomethane production from wheat plant under mesophilic anaerobic digestion. At 121°C, they obtained a maximum methane yield of 15.5% higher than that of the untreated wheat plant after pretreatment for 120 min.

4.3.2 Alkali

The mechanism of alkaline pretreatment mainly induces swelling of particulate organics at elevated pH, enabling the biomass cellular substances more susceptible to enzymatic action [24]. The complex cell gets damaged by the hydroxyl anions available in the alkali. In macroalgae, it enhances hydrolysis of RNA, organic liquefaction of proteins and saponification [28]. In lignocellulosic biomass, it causes swelling, delignification and de-esterification of intermolecular ester bonds. With the disintegration of the bonds the porosity and internal surface area of the biomass increases, the degree of polymerization and crystallinity decreases. This makes it more accessible for enzymes and bacteria [6]. Regarding WAS, at higher pH, the microbial cell walls are broken and intracellular material is released into the liquid phase.

Studies were explored by Banu et al. [13] to evaluate the advantage of sodium hydroxide (NaOH) for its higher sludge solubilization potential and lime. They conducted experiments at a fixed alkali strength (35 meq/l) and varying concentration of NaOH and lime to demonstrate the role of alkalis in solubilizing sludge. The highest solubilization they achieved, was at an optimum dosage of NaOH and lime 1.6 and 0.7 g/l respectively at time 3 h. Sambusiti et al. [75] investigated the effect of alkaline (NaOH) pretreatment on ensiled sorghum forage in semi continuous digesters. They observed that pretreatment with 10 g NaOH/100 g TS increased the methane yield by 25% compared to untreated sorghum without experiencing any inhibition of the process.

4.3.3 Oxidative

Wet air oxidation is a pretreatment option that enhances contact between molecular oxygen and organic matter for the complete degradation of organic compounds into carbon dioxide and water. In order to achieve this, high temperature (and subsequently high pressure) conditions are required [22]. The correspondingly high pressure required is to maintain the high temperature conditions, as well as to help increase the concentration of dissolved oxygen, and thus oxidation rate. Chandra et al. [76] employed wet air oxidation to enhance the biodegradability of the complex biomethanated distillery effluent. They reported an enhanced biogas yield of pretreated effluent up to 2.8 times higher than the untreated effluent with methane content up to 64.14%.

4.3.4 Ozonation

Ozone is a strong oxidant and hence powerful in oxidizing substrates. It has potential to degrade lignin in diverse feedstocks. It reacts with the polysaccharides, proteins, lipids and other recalcitrant compounds and transform them into biodegradable molecules. The ozonation process can result in efficient cell wall rupture and release of more soluble and easily biodegradable organics, which can be easily accessed and assimilated by anaerobic microorganisms. Thus it leads to improvement in the AD process [15].

AD of ozone pretreated excess sludge was studied by Goel et al. [14] through long-term operation of laboratory-scale reactors. They found that ozone pretreatment was effective in partially solubilizing the sludge solids and leading to subsequent improvement in anaerobic degradability. The extent of solubilization and digestion efficiency depended on the applied ozone doses. At 0.05 g O_3/g TS, the AD efficiencies improved to about 59% as compared to 31% for the control run. Different process indicators like specific methane production and ammonia concentration in the reactor, also specify the higher observed solid degradation rates for ozonated sludge.

4.4 Biological

The biological mediated pretreatment process is based on the function of multiple form of heterotrophic microbes. Complex biopolymers such as protein and carbohydrate can be transformed into simpler end products due to the action of various enzymes produced by the bacteria. The significance of biological pretreatment lies in the fact that is solubilizes the organic compounds present in the biomass with minimum energy, with no severe changes in substrate environment. Biological pretreatment is done with or without enzyme addition some of which can be produced endogenously by microorganisms present in the sludge. Some of the enzymes like protease, lipase, cellulase, alpha-amylase and dextranase [11] can effectively improve the hydrolysis rate and release of biopolymers to a large extent. Contrarily, these enzymes are more costly and difficult to preserve. Bonilla et al. [77] evaluated the potential for enzymatic pretreatment of pulp mill biosludge with protease from *B. licheniformis* for biodegradability. Carrying out BMP test, they arrived at a maximum improvement of 26% in biogas yield.

Saranya et al. [10] studied the impacts of phase separated disintegration pretreatment using calcium chloride ($CaCl_2$) and bacteria. For their study a pH of 6.5, temperature of 40°C and treatment period of 42 h were the optimum conditions for pretreatment. In the initial phase, they achieved the floc disruption (deflocculation) with 0.06 g/g SS of $CaCl_2$ and in the latter phase, cell disintegration through potent biosurfactant producing bacteria, *Planococcus jake 01*. They were able to achieve 17.14% SS reduction and 14.14% COD solubilization for deflocculated and bacterially pretreated sludge, which were comparatively higher than for sludge treated with bacteria alone. They observed a biogas yield potential for pretreated sludge of 0.322 L/g VS as against 0.145 L/g VS for control. Kavitha et al. [43] investigated the bacterial-based biological pretreatment on liquefaction of microalga *Chlorella vulgaris* with cellulase-secreting bacteria prior to anaerobic biodegradation. The biomethanation studies implied that bacterial pretreatment increased the bioavailability of biomass and hence methane generation. They arrived at a methane yield of nearly twice that of control.

Fungal pretreatment improves degradation of lignin and hemicellulose and hence result in increased digestibility of cellulose, which is preferably essential for AD process. Several fungal classes, including brown-, white- and soft-rot fungi, have been used for pretreatment of lignocellulosic biomass for biogas production, with white-rot fungi being the most effective. Amirta et al. [78] employed four fungal species to pretreat Japanese cedar wood chips in the presence of wheat bran which supplements nutrition for fungal growth. They revealed that wood chips pretreated by *Ceriporiopsis subvermispora ATCC 90467* produced the highest methane yield, which was 4 times higher than that of the control biomass at the end of 8 weeks.

4.5 Combined treatment

4.5.1 Steam explosion

Steam explosion pretreatment is an effort to expose the biomass to high temperature and pressure for short period of time and then reducing the pressure rapidly. This stops the reactions, causing the biomass to decompose explosively. This pretreatment condition may involve temperatures as high as 260°C and pressure up to 4.5 MPa. A study was investigated by Nges et al. [79] to improve the anaerobic biodegradability of *Miscanthus lutarioriparius* for biogas production. Employing steam explosion pretreatment with 0.3 M NaOH with particle size reduced to 0.5, they achieved a methane yield of 57% higher than that for the untreated samples. Their result was estimated to be 71% of theoretical methane yield of the biomass. Wang et al. [80] achieved a 24% higher methane yield than untreated bulrush at 1.72 MPa steam pressure, 8.14 min residence time, and 11% moisture content employing steam-explosion treatment of bulrush. During the pretreatment they observed the breakage, disruption, and redistribution of the rigid lignin structure which was proved by thermos gravimetric analysis. Srisang and Chavalparit [81] optimized a pre-treatment condition of 1.0% acetic acid, 17.45 min reaction time of sugarcane bagasse using steam explosion at 180° C. They achieved a maximum biogas production (434.47 L/kg VS) which was 91.88% higher than that of control (226.42 L/kg VS).

4.5.2 Physico chemical

The combination of thermal and chemical pre-treatments have been investigated in a number of studies in which the enhancement of the anaerobic digestibility of sludge was reported. Yi et al. [19] has used combined alkaline and low-temperature thermal pretreatment to enhance the subsequent AD of WAS. Different combinations of these two methods were investigated and biochemical methane potential (BMP) test was used to assess the anaerobic digestibility of pretreated WAS. With the combined treatment of adding 0.05 g NaOH/g TS and temperature maintained at 70°C for 9 h, they achieved a ratio of 72.8% soluble carbohydrate/total carbohydrate. Biogas production achieved through their BMP experiment was six times higher than the control and the average value of methane content of the produced biogas was 64%. In another study, Kavitha et al. [56], employed microwave irradiation to disintegrate the dairy WAS biomass after deagglomerating the sludge using a mechanical device, ultrasonicator. The outcomes of their study revealed that a higher biomass lysis efficiency of about 33.2% was possible through ultrasonic assisted microwave disintegration (UMWD) when compared to microwave disintegration MWD (20.9%). Their results of BMP test showed that UMWD has better amenability towards AD with 50% higher methane production representing enhanced liquefaction potential of disaggregated sludge biomass.

Jang and Ahn [5] determined the effect of MW irradiation with NaOH pretreatment on AD of thickened municipal WAS in semi-continuous mesophilic digesters at HRT of 15, 10, 7, and 5 days. They combined MW pretreatment at temperature of 135°C with the input power of 1000 W with 60 ml of alkaline (20 meq NaOH/l) pretreated sludge. The degree of substrate solubilization arrived was 18 times higher in pretreated sludge (53.2%) than in raw sludge (3.0%). With HRT reduced to 5 days, they observed an improvement in biogas production (205% higher) for pretreated sludge compared with the control. The results show that MW irradiation combined with alkali pretreatment is effective in increasing mesophilic anaerobic biodegradability of sewage sludge. Ebenezer et al. [58] reported an increased COD and biopolymers release of WAS treated with Sodium citrate, a cationic binding agent, followed by microwaves pretreatment. They also concluded that the above pretreatment made the biomass more amenable for batch AD and hence higher biogas production with a methane content of 60–70% of biogas volume. Tamilarasan et al. [28] has made an attempt, by coupling a mechanical disperser with a chemical Sodium tripolyphosphate (STPP) for pretreatment of macroalgal biomass. They arrived at a 15% liquefaction and more than 5 times higher methane production compared to control at an optimal disperser-specific energy input of about 3312.67 kJ/kg TCOD (total COD) and an STPP dosage of about 0.04 g/g COD. Thus the combined pretreatment showed a greater biodegradability and biomethanation properties.

4.5.3 Ammonia fiber expansion

Ammonia fiber expansion is a promising method especially to pretreat agricultural materials for bioenergy production. Ammonia can be easily recovered and presents a high selectivity towards the lignin reactions, while preserving the carbohydrates. Ammonia can also penetrate the crystalline structure of cellulose and causes swelling [30]. The method involves treating the lignocellulosic biomass with liquid ammonia under mild temperature (70–200°C) and pressure (100–400 psi) for a specific time. This explosion results in several physical and chemical alterations in the structure of biomass. Jurado et al. [32] studied the effect of aqueous ammonia soaking (AAS) as a method to disrupt the lignocellulosic structure and increase the methane yield of wheat straw, miscanthus and willow. In all three cases, with AAS they observed an increase in methane yield from 37 to 41%, 25 to 27% and 94 to 162% for wheat straw, miscanthus and willow, respectively. Antonopoulou et al. [30] employed AAS as a pretreatment method, for the AD of three lignocellulosic biomass—poplar sawdust, sunflower straw and grass. In their study, they arrived at an increase in the ultimate methane yield being 148.7, 37.7 and 26.2% of poplar, sunflower straw and grass, respectively. They did not observe any toxic compounds such as furaldehydes, during AAS pretreatment.

4.6 Electrical

In this pretreatment, a very short burst (~100 μs) of rapidly pulsed (several kHz), high voltage (about 20 kV) electric field is utilized to disrupt and break up the cell membrane of microorganism. This focused pulse (FP) induces a critical electrical potential across the cell membrane, causing cell lysis by direct attack on phospholipids and the peptidoglycan, respectively. Once the cell membranes get damaged, the intracellular organic material are released, making complex organic macromolecules more biodegradable [82]. They evaluated the effects of FP treatment and SRT on WAS in laboratory-scale digesters operated at SRTs of 2–20 days. They achieved an increased methane production rate and TCOD removal efficiency of about 33% and 18%, respectively, at a SRT of 20 days. They also concluded that, an increase in the hydrolysis rate was caused by FP-treatment of WAS, particularly at lower SRTs. Salerno et al. [83] applied FP to WAS and pig manure for increasing the production of methane during AD. In their work, methane production increased 200% for sludge and 80% for pig manure as compared to untreated sludge and manure. Thus PEF technology is advantageous due to low energy requirement for very short pulse time.

5. Future challenges and conclusion

The global energy supply is highly relying on fossil sources (crude oil, coal, natural gas) till now. According to the current energy policies and management, world market energy consumption is forecast to increase by 44% from 2006 to 2030 [84]. At the same time, concentrations of greenhouse gases in the atmosphere are rising rapidly, with fossil fuel-derived CO_2 emissions being the most important contributor. Nowadays, increasing attention has been gained on various strategies for the bioconversion of biomass into methane-rich biogas, due to increased global warming, the need for sustainable waste management and high energy costs [41]. The production of biogas through AD offers significant advantages over other forms of bioenergy production. Unlike fossil fuels, biogas from AD is permanently renewable, as it is produced from biomass, which is a living form of storage of solar energy through photosynthesis [85]. It has been evaluated as one of the most energy-efficient and environmentally beneficial technology for bioenergy production [86]. It can drastically reduce GHG emissions compared to fossil fuels by utilization of locally available resources.

Many sources, such as crops, grasses, leaves, manure, fruit, and vegetable wastes or algae can be used, and the process can be applied in small and large scales in many parts of the world. Energy crops digestion requires prolonged HRT of several weeks to month to achieve complete fermentation with high gas yields and minimized residual gas potential of the digestate [4]. For an increased dissemination of biogas plants, further improvements of the process efficiency, and the development of new technologies for mixing, process monitoring, and process control are necessary. Pretreatment of substrates and the addition of micronutrients offers a major potential for increasing the biogas yield. With the increasing number of biogas plants, also an improvement of the effluent quality is necessary, in order to avoid a contamination of ground water with pathogens and nutrients [3]. The choice of a pretreatment should be made not only based on energy balance and economy, but also various environmental factors such as pathogen removal, use of chemicals, and the possibility for a sustainable use of the residues, impacts on human health and the environment [8]. Carballa et al. [87] evaluated the environmental aspects of different pretreatment methods in terms of abiotic resources depletion potential, eutrophication potential, global warming potential, human and terrestrial toxicity potential through a life cycle assessment.

The profitable operation of a biogas plant relies on low capital and operational expenditures [28]. The frequent approaches including physical, thermal and chemical processes have been commercially implemented nowadays with a number of patented technologies. But research on biological techniques is still undergoing investigations from bench scale to full scale applications. Many pretreatment methods are expensive or have a high energy demand. The performance of any pretreatment method is quantified based on the economic feasibility of the method in terms of the cost of pretreatment versus the value of added methane yield. The effect of the pretreatment is however mostly dependent on the biomass composition and operating conditions. The investment costs for pretreatment of recalcitrant substrates are high at the moment due to high expenditure in process engineering. Biological disintegration is devoid of chemical contamination and energy inputs and the use of an enzyme secreting bacterial consortium for biomass is beneficial, as commercial enzymes are expensive [55]. But the need for long reaction times renders biological pretreatment unsuitable for large scale plants where land space is expensive or restricted.

Most studies reviewed assessed the impact of pretreatment processes on the biogas yield on a laboratory scale with a few determining the net energy gain/loss obtained after pretreatment [11, 28, 58]. Most studies in the literature are conducted as lab scale experiments and do not represent the same output that could be achieved through large scale biogas production facilities. Hence, there is a continuous need for newer and cleaner methods of biomass processing with less energy demand and lower waste generation.

This chapter concludes the effect of various biomass pretreatment for enhancement of biogas production and the future challenges for an energy efficient and eco-friendly manner. Therefore, optimizing the pretreatment conditions in order to lower production costs, improving the process performance and production of fewer residues is needed. A pretreatment method optimized based on the above situations may enhance the performance of individual pretreatments and achieve technical, environmental and financial feasibility. However, a further research on combined pretreatments is necessary in the future to get useful information that may lead to the necessary improvements in the AD industry.

Author details

Tamilarasan Karuppiah[1] and Vimala Ebenezer Azariah[2]*

1 Department of Civil Engineering, Regional Centre of Anna University, Tirunelveli, India

2 V V College of Engineering, Tirunelveli, India

*Address all correspondence to: vimalajulius1@gmail.com

References

[1] Kannah RY, Kavitha S, Banu JR, Karthikeyan OP, Sivashanmugham OP. Dispersion induced ozone pretreatment of waste activated biosolids: Arriving biomethanation modelling parameters, energetic and cost assessment. Bioresource Technology. 2017;**244**: 679-687. DOI: 10.1016/j.biortech. 2017.08.001

[2] Kuglarz M, Karakashev D, Angelidaki I. Microwave and thermal pretreatment as methods for increasing the biogas potential of secondary sludge from municipal wastewater treatment plants. Bioresource Technology. 2013; **134**:290-297. DOI: 10.1016/j.biortech. 2013.02.001

[3] Weiland P. Biogas production: Current state and perspectives. Applied Microbiology and Biotechnology. 2010; **85**:849-860. DOI: 10.1007/s00253-009-2246-7

[4] Kuusik A, Pachel K, Loigu E. Possible agricultural use of digestate. Proceedings of the Estonian Academy of Sciences. 2017;**66**(1):64-74. DOI: 10.3176/proc.2017.1.10

[5] Jang J, Ahn J. Effect of microwave pretreatment in presence of NaOH on mesophilic anaerobic digestion of thickened waste activated sludge. Bioresource Technology. 2013;**131**: 437-442. DOI: 10.1016/j.biortech. 2012.09.057

[6] Patinvoh RJ, Osadolor OA, Chandolias K, Horváth IS, Taherzadeh MJ. Innovative pretreatment strategies for biogas production. Bioresource Technology. 2016;**224**:13-24. DOI: 10.1016/j.biortech.2016.11.083

[7] Apul OG, Sanin FD. Ultrasonic pretreatment and subsequent anaerobic digestion under different operational conditions. Bioresource Technology. 2010;**101**:8984-8992. DOI: 10.1016/j. biortech.2010.06.128

[8] Salsabil MR, Prorot A, Casellasa M, Dagot C. Pretreatment of activated sludge: Effect of sonication on aerobic and anaerobic digestibility. Chemical Engineering Journal. 2009;**148**:327-335. DOI: 10.1016/j.cej.2008.09.003

[9] Show KY, Mao T, Lee DJ. Optimization of sludge disruption by sonication. Water Research. 2007;**41**: 4741-4747. DOI: 10.1016/j.watres. 2007.07.017

[10] Saranya T, Kavitha S, Kaliappan S, Adish Kumar S, Yeom IT, Banu JR. Accelerating the sludge disintegration potential of a novel bacterial strain Planococcus jake 01 by $CaCl_2$ induced deflocculation. Bioresource Technology. 2015;**175**:396-405. DOI: 10.1016/j. biortech.2014.10.122

[11] Kavitha S, Adish Kumar S, Yogalakshmi KN, Kaliappan S, Banu JR. Effect of enzyme secreting bacterial pretreatment on enhancement of aerobic digestion potential of waste activated sludge interceded through EDTA. Bioresource Technology. 2013; **150**:210-219. DOI: 10.1016/j.biortech. 2013.10.021

[12] Merrylin J, Kaliappan S, Adish Kumar S, Yeom IT, Banu JR. Enhancing aerobic digestion potential of municipal waste-activated sludge through removal of extracellular polymeric substance. Environmental Science and Pollution Research International. 2013;**21**(2): 1112-1123. DOI: 10.1007/s11356-013-1976-3

[13] Banu JR, Khac UD, Kumar SA, Yeom IT, Kaliappan S. A novel method of sludge pretreatment using the combination of alkalis. Journal of Environmental Biology. 2012;**33**:249-253

[14] Goel R, Tokutomi T, Yasui H. Anaerobic digestion of excess activated sludge with ozone pretreatment. Water Science and Technology. 2013; **47**:207-214. DOI: 10.2166/wst.2003.0648

[15] Elliott A, Mahmood T. Pretreatment technologies for advancing anaerobic digestion of pulp and paper biotreatment residues. Water Research. 2007;**41**:4273-4286. DOI: 10.1016/j.watres.2007.06.017

[16] Beszédes S, László Z, Horváth ZH, Szabó G, Hodúr C. Comparison of the effects of microwave irradiation with different intensities on the biodegradability of sludge from the dairy and meat industry. Bioresource Technology. 2011;**102**:814-821. DOI: 10.1016/j.biortech.2010.08.121

[17] Eskicioglu C, Kennedy KJ, Droste RL. Enhancement of batch waste activated sludge digestion by microwave pretreatment. Water Environment Research. 2007;**79**:2304-2317. DOI: 10.2175/106143007X184069

[18] Borges ESM, Chernicharo CAL. Effect of thermal treatment of anaerobic sludge on the bioavailability and biodegradability characteristics of the organic fraction. Brazilian Journal of Chemical Engineering. 2009;**26**: 469-480. DOI: 10.1590/S0104-66322009000300003

[19] Yi H, Han Y, Zhuo Y. Effect of combined pretreatment of waste activated sludge for AD process. Procedia Environmental Sciences. 2013; **18**:716-721. DOI: 10.1016/j.proenv.2013.04.097

[20] Sahinkaya S, Sevimli MF. Sono-thermal pre-treatment of waste activated sludge before anaerobic digestion. Ultrasonics Sonochemistry. 2013;**20**:587-594. DOI: 10.1016/j.ultsonch.2012.07.006

[21] Rani UR, Kaliappan S, Adish Kumar AS, Banu JR. Combined treatment of alkaline and disperser for improving solubilisation and anaerobic biodegradability of dairy waste activated sludge. Bioresource Technology. 2012;**126**:107-116. DOI: 10.1016/j.biortech.2012.09.027

[22] Strong PJ, McDonald B, Gapes DJ. Combined thermo alkaline and fermentative destruction of municipal biosolids: A comparison between thermal hydrolysis and wet oxidative pre-treatment. Bioresource Technology. 2011;**102**:5520-5527. DOI: 10.1016/j.biortech.2010.12.027

[23] Shah FA, Mahmoo Q, Shah MM, Pervez A, Asad SA. Microbial ecology of anaerobic digesters: The key players of anaerobiosis. Scientific World Journal. Article ID 183752. 2014. Open Access. 21 p. DOI: 10.1155/2014/183752

[24] Gumisiriza R, Hawumba JF, Okure M, Hensel O. Biomass waste-to-energy valorization technologies: A review case for banana processing in Uganda. Biotechnology for Biofuels. 2017;**10**:11. DOI: 10.1186/s13068-016-0689-5

[25] Mir MA, Hussain A, Verma C. Design considerations and operational performance of anaerobic digester: A review. Cogent Engineering. 2016;**3**: 1181-1196. DOI: 10.1080/23311916.2016.1181696

[26] Metcalf E. Wastewater Engineering: Treatment and Reuse. 4th ed. New York: McGraw-Hill; 2003

[27] Verma. Anaerobic digestion of biodegradable organics in municipal solid wastes [thesis]. In: Fu Foundation School of Engineering & Applied Science Columbia University. 2002

[28] Tamilarasan K, Kavitha S, Banu JR, Arulazhagan P, Yeom IT. Energy-efficient methane production from

macroalgal biomass through chemo disperser liquefaction. Bioresource Technology. 2017;**228**:156-163. DOI: 10.1016/j.biortech.2016.12.102

[29] Gayathri T, Kavitha S, Adish Kumar S, Kaliappan S, Yeom IT, Banu JR. Effect of citric acid induced deflocculation on the ultrasonic pretreatment efficiency of dairy waste activated sludge. Ultrasonics Sonochemistry. 2015;**22**: 333-340. DOI: 10.1016/j.ultsonch. 2014.07.017

[30] Antonopoulou G, Gavala HN, Skiadas IV, Lyberatos G. The effect of aqueous ammonia soaking pretreatment on methane generation using different lignocellulosic biomasses. Waste and Biomass Valorization. 2015; **6**:281-291. DOI: 10.1007/s12649-015-9352-9

[31] Lymperatou A, Gavala HN, Esbensen KH, Skiadas IV. AMMONOX: Ammonia for enhancing biogas yield and reducing nox—Analysis of effects of aqueous ammonia soaking on manure fibers. Waste and Biomass Valorization. 2015;**6**:449-457. DOI: 10.1007/ s12649-015-9365-4

[32] Jurado E, Gavala HN, Skiadas IV. Enhancement of methane yield from wheat straw, miscanthus and willow using aqueous ammonia soaking. Environmental Technology. 2013;**34**: 2069-2075. DOI: 10.1080/09593330. 2013.826701

[33] Rani UR, Adish Kumar S, Kaliappan S, Yeom IT, Banu J. Impacts of microwave pre-treatments on the semi-continuous anaerobic digestion of dairy waste activated sludge. Waste Management. 2013;**33**:1119-1127. DOI: 10.1016/j.wasman.2013.01.016

[34] Teghammar A, Karimi K, Horva'th IS, Mohammad Taherzadeh J. Enhanced biogas production from rice straw, triticale straw and softwood spruce by NMMO pretreatment. Biomass and Bioenergy. 2012;**36**:116-120. DOI: 10.1016/j.biombioe.2011.10.019

[35] Lee HV, Hamid SBA, Zain SK. Conversion of lignocellulosic biomass to nanocellulose: Structure and chemical process. The Scientific World Journal. Article ID 631013. 2014. Open Access. 20 p. DOI: 10.1155/2014/ 631013

[36] Daroit DJ, Corrêa APF, Brandelli A. Keratinolytic potential of a novel *Bacillus sp.* P45 isolated from the amazon basin fish piaractus mesopotamicus. International Biodeterioration & Biodegradation. 2009;**63**:358-363. DOI: 10.1016/j. ibiod.2008.11.008

[37] Barone JR, Schmidt WF, Gregoire NT. Extrusion of feather keratin. Journal of Applied Polymer Science. 2006; **100**(2):1432-1442. DOI: 10.1002/ app.23501

[38] Ebenezer AV, Kaliappan S, Adish Kumar S, Yeom IT, Banu JR. Influence of deflocculation on microwave disintegration and anaerobic biodegradability of waste activated sludge. Bioresource Technology. 2015; **185**:194-201. DOI: 10.1016/j. biortech.2015.02.102

[39] Kim HT, Yun EJ, Wang D, Chung JH, Choi IG, Kim KH. A high temperature and low acid pretreatment and agarose treatment of agarose for the production of sugar and ethanol from red seaweed biomass. Bioresource Technology. 2013;**136**: 582-587. DOI: 10.1016/j.biortech. 2013.03.038

[40] Motte JC, Escudié R, Hamelin J, Steyer JP, Bernet N, Delgenes JP, et al. Substrate milling pretreatment as a key parameter for solid-state anaerobic digestion optimization. Bioresource Technology. 2014;**173**:185-192. DOI: 10.1016/j.biortech.2014.09.015

[41] Li C, Liu G, Nges IA, Liu J. Enhanced biomethane production from *Miscanthus lutarioriparius* using steam explosion pretreatment. Fuel. 2016;**179**: 267-273. DOI: 10.1016/j.fuel.2016. 03.087

[42] Zeynali R, Khojastehpour M, Ebrahimi-Nik M. Effect of ultrasonic pre-treatment on biogas yield and specific energy in anaerobic digestion of fruit and vegetable wholesale market wastes. Sustainable Environment Research. 2017;**27**:259-264. DOI: 10.1016/j.serj.2017.07.001

[43] Kavitha S, Subbulakshmi P, Banu JR, Gopi M, Yeom IT. Enhancement of biogas production from microalgal biomass through cellulolytic bacterial pretreatment. Bioresource Technology. 2017;**233**:34-43. DOI: 10.1016/j. biortech.2017.02.081

[44] Kavitha S, Jayashree C, Adish Kumar S, Kaliappan S, Banu JR. Enhancing the functional and economical efficiency of a novel combined thermo chemical disperser disintegration of waste activated sludge for biogas production. Bioresource Technology. 2014;**173**:32-41. DOI: doi. org/10.1016/j.biortech.2014.09.078

[45] Poornima Devi T, Ebenezer AV, Adish Kumar S, Kaliappan S, Banu JR. Effect of deflocculation on the efficiency of disperser induced dairy waste activated sludge disintegration and treatment cost. Bioresource Technology. 2014;**167**:151-158. DOI: 10.1016/j.biortech.2014.06.004

[46] Rani RU, Adish Kumar S, Kaliappan S, Banu JR. Enhancing the anaerobic digestion potential of dairy waste activated sludge by two step sono-alkalization pretreatment. Ultrasonics Sonochemistry. 2014;**21**:1065-1074. DOI: 10.1016/j.ultsonch.2013.11.007

[47] Kavitha S, Kannah RY, Yeom IT, Do KU, Banu JR. Combined thermo-chemo-sonic disintegration of waste activated sludge for biogas production. Bioresource Technology. 2015;**197**:383-392. DOI: 10.1016/j. biortech.2015.08.131

[48] Kavitha S, Banu JR, IvinShaju CD, Kaliappan S, Yeom IT. Fenton mediated ultrasonic disintegration of sludge biomass: Biodegradability studies, energetic assessment, and its economic viability. Bioresource Technology. 2016; **221**:1-8. DOI: 10.1016/j.biortech.2016. 09.012

[49] Kavitha S, Banu JR, Subitha G, Ushani U, Yeom IT. Impact of thermo-chemo-sonic pretreatment in solubilizing waste activated sludge for biogas production: Energetic analysis and economic assessment. Bioresource Technology. 2016;**219**:479-486. DOI: 10.1016/j.biortech.2016.07.115

[50] Kavitha S, Banu JR, Kumar JV, Rajkumar M. Improving the biogas production performance of municipal waste activated sludge via disperser induced microwave disintegration. Bioresource Technology. 2016;**217**:21-27. DOI: 10.1016/j.biortech.2016.02.034

[51] Kavitha S, Pray SS, Yogalakshmi KN, Adish Kumar S, Yeom IT, Banu JR. Effect of chemo-mechanical disintegration on sludge anaerobic digestion for enhanced biogas production. Environmental Science and Pollution Research. 2016;**23**:2402-2414. DOI: 10.1007/s11356-015-5461-z

[52] Kavitha S, Brindha GMJ, Gloriana AS, Rajashankar K, Yeom IT, Banu JR. Enhancement of aerobic biodegrdability potential of municipal waste activated sludge by ultrasonic aided bacterial disintegration. Bioresource Technology. 2016;**200**:161-169. DOI: 10.1016/j. biortech.2015.10.026

[53] Ushani U, Banu JR, Kavitha S, Kaliappan S, Yeom IT. Immobilized and $MgSO_4$ induced cost effective bacterial

disintegration of waste activated sludge for effective anaerobic digestion. Chemosphere. 2017;**175**:66-75. DOI: 10.1016/j.chemosphere.2017.02.046

[54] Shanthi M, Banu JR, Sivashanmugham P. Effect of surfactant assisted sonic pretreatment on liquefaction of fruits and vegetables residue: Characterization, acidogenesis, biomethane yield and energy ratio. Bioresource Technology. 2018;**264**: 35-41. DOI: 10.1016/j.biortech.2018.05.054

[55] Banu JR, Kannah RK, Kavitha S, Gunasekaran M, Yeom IT, Kumar G. Disperser-induced bacterial disintegration of partially digested anaerobic sludge for efficient biomethane recovery. Chemical Engineering Journal. 2018;**347**:165-172. DOI: 10.1016/j.cej.2018.04.096

[56] Kavitha S, Banu JR, Kumar G, Kaliappan S, Yeom IT. Profitable ultrasonic assisted microwave disintegration of sludge biomass: Modelling of biomethanation and energy parameter analysis. Bioresource Technology. 2018;**254**:203-213. DOI: 10.1016/j.biortech.2018.01.072

[57] Tamilarasan K, Arulazhagan P, Uma Rani R, Kaliappan S, Banu JR. Synergistic impact of sonic—Tenside on biomass disintegration potential: Acidogenic and methane potential studies, kinetics and cost analytics. Bioresource Technology. 2018;**253**: 256-261. DOI: 10.1016/j.biortech.2018.01.028

[58] Ebenezer AV, Arulazhagan P, Adish Kumar S, Yeom IT, Banu JR. Effect of deflocculation on the efficiency of low energy microwave pretreatment and anaerobic biodegradability of waste activated sludge. Applied Energy. 2015; **145**:104-110. DOI: 10.1016/j.apenergy.2015.01.133

[59] Eswari P, Kavitha S, Kaliappan S, Yeom IT, Banu JR. Enhancement of sludge anaerobic biodegradability by combined microwave-H_2O_2 pretreatment in acidic conditions. Environmental Science and Pollution Research. 2016;**23**:13467-13479. DOI: 10.1007/s11356-016-6543-2

[60] Eswari AP, Kavitha S, Banu JR, Parthiba Karthikeyan O, Yeom IT. H_2O_2 induced cost effective microwave disintegration of dairy waste activated sludge in acidic environment for efficient biomethane generation. Bioresource Technology. 2017;**244**: 688-697. DOI: 10.1016/j.biortech.2017.07.078

[61] Zheng Y, Zhao J, Xu F, Li Y. Pretreatment of lignocellulosic biomass for enhanced biogas production. Progress in Energy and Combustion Science. 2014;**42**:35-53. DOI: 10.1016/j.pecs.2014.01.001

[62] Dahadha S, Amin Z, Lakeh AAB, Elbeshbishy E. Evaluation of different pretreatment processes of lignocellulosic biomass for enhanced biomethane production. Energy & Fuels. 2017;**31**(10):10335-10347. DOI: 10.1021/acs.energyfuels.7b02045

[63] Montgomery LFR, Bochmann G. Pretreatment of feedstock for enhanced biogas production. IEA Bioenergy. Technical Brochure. 2014:**22**

[64] Gogate PR, Sutkar VS, Pandit AB. Sonochemical reactors: Important design and scale up considerations with a special emphasis on heterogeneous systems. Chemical Engineering Journal. 2011;**166**:1066-1082. DOI: 10.1016/j.cej.2010.11.069

[65] Alzate ME, Munoz R, Rogalla F, Fdz-Polanco F, Perez-Elvira SI. Biochemical methane potential of microalgae: Influence of substrate to inoculum ratio, biomass concentration

and pretreatment. Bioresource Technology. 2012;**123**:488-494. DOI: 10.1016/j.biortech.2012.06.113

[66] Patil PN, Gogate PR, Csoka L, Dregelyi-Kiss A, Horvath M. Intensification of biogas production using pretreatment based on hydrodynamic cavitation. Ultrasonics Sonochemistry. 2016;**30**:79-86. DOI: 10.1016/j.ultsonch.2015.11.009

[67] Rincón B, Bujalance L, Fermoso FG, Martín A, Borja R. Biochemical methane potential of two-phase olive mill solid waste: Influence of thermal pretreatment on the process kinetics. Bioresource Technology. 2013;**140**: 249-255. DOI: 10.1016/j.biortech.2013.04.090

[68] Maroušek J. Finding the optimal parameters for the steam explosion process of hay. Revista Técnica de la Facultad de Ingeniería Universidad del Zulia. 2012;**35**(2):170-178. DOI: 10.1186/s40643-017-0137-9

[69] Novarino D, Zanetti MC. Anaerobic digestion of extruded OFMSW. Bioresource Technology. 2011;**104**: 44-50. DOI: 10.1016/j.biortech.2011.10.001

[70] Wilson C, Novak JT. Hydrolysis of macromolecular components of primary and secondary wastewater sludge by thermal hydrolytic pretreatment. Water Research. 2009;**43**:4489-4498. DOI: 10.1016/j.watres.2009.07.022

[71] Jayashree C, Janshi G, Yeom IT, Kumar SA, Banu JR. Effect of low temperature thermo-chemical pretreatment of dairy waste activated sludge on the performance of microbial fuel cell. International Journal of Electrochemical Science. 2014;**9**: 5732-5742

[72] Li C, Knierim B, Manisseri C, Arora R, Scheller HV, Auer M, et al. Comparison of dilute acid and ionic liquid pretreatment of switchgrass: Biomass recalcitrance, delignification and enzymatic saccharification. Bioresource Technology. 2010;**101**: 4900-4906. DOI: 10.1016/j.biortech.2009.10.066

[73] Devlin DC, Esteves SRR, Dinsdale RM, Guwy AJ. The effect of acid pretreatment on the anaerobic digestion and dewatering of waste activated sludge. Bioresource Technology. 2011; **102**:4076-4082. DOI: 10.1016/j.biortech.2010.12.043

[74] Taherdanak M, Zilouei H, Karimi K. The influence of dilute sulfuric acid pretreatment on biogas production form wheat plant. International Journal of Green Energy. 2016;**13**(11): 1129-1134. DOI: 10.1080/15435075.2016.1175356

[75] Sambusiti C, Ficara E, Malpei F, Steyer JP, Carrère H. Benefit of sodium hydroxide pretreatment of ensiled sorghum forage on the anaerobic reactor stability and methane production. Bioresource Technology. 2013;**144**: 149-155. DOI: 10.1016/j.biortech.2013.06.095

[76] Chandra TS, Malik SN, Suvidha G, Padmere ML, Shanmugam P, Mudliar SN. Wet air oxidation pretreatment of biomethanated distillery effluent: Mapping pretreatment efficiency in terms color, toxicity reduction and biogas generation. Bioresource Technology. 2014;**158**:135-140. DOI: 10.1016/j.biortech.2014.01.106

[77] Bonilla S, Choolaei Z, Meyer T, Edwards EA, Yakunin AF, Allen DG. Evaluating the effect of enzymatic pretreatment on the anaerobic digestibility of pulp and paper biosludge. Biotechnology Reports. 2018; **17**:77-85. DOI: doi.org/10.1016/j.btre.2017.12.009

[78] Amirta R, Tanabe T, Hondaa Y, Kuwahara M, Watanabe T. Methane fermentation of Japanese cedar wood pretreated with a white rot fungus, *Ceriporiopsis subvermispora*. Journal of Biotechnology. 2006;**123**:71-77. DOI: 10.1016/j.jbiotec.2005. 10.004

[79] Nges IA, Li C, Wang B, Xiao L, Yi Z, Liu J. Physio-chemical pretreatments for improved methane potential of *Miscanthus lutarioriparius*. Fuel. 2016; **166**:29-35. DOI: 10.1016/j.fuel.2015. 10.108

[80] Wang J, Yue ZB, Chen TH, Peng SC, Yu HQ, Chen HZ. Anaerobic digestibility and fiber composition of bulrush in response to steam explosion. Bioresource Technology. 2010;**101**: 6610-6614. DOI: 10.1016/j. biortech.2010.03.086

[81] Srisang N, Chavalparit O. Enhancing biogas production from sugarcane bagasse using steam explosion in according with acetic acid pretreatment. Advanced Materials Research. 2013;**856**:321-326. DOI: 10.4028/www.scientific.net/ AMR.856.321

[82] Lee I, Rittmann BR. Effect of low solids retention time and focused pulsed pre-treatment on AD of waste activated sludge. Bioresource Technology. 2011; **102**:2542-2548. DOI: 10.1016/j. biortech.2010.11.082

[83] Salerno MB, Lee HS, Parameswaran P, Rittmann B. Using a pulsed electric field as a pretreatment for improved biosolids digestion and methanogenesis. Water Environment Research. 2009; **81**(8):831-839. DOI: 10.2175/ 106143009X407366

[84] IEO, International Energy Outlook. Energy Information Administration. Office of Integrated Analysis and Forecasting. Washington, DC: US Department of Energy; 2009. p. 284

[85] Al Seadi T, Rutz D, Prassl H, Köttner M, Finsterwalder T, Volk S, et al. Biogas Handbook. Esbjerg, Denmark: University of Southern Denmark Esbjerg, Niels Bohrs Vej; 2008. pp. 9-10

[86] Panwar NL, Kaushik SC, Kothari S. Role of renewable energy sources in environmental protection: A review. Renewable and Sustainable Energy Reviews. 2011;**15**:1513-1524. DOI: 10.1016/j.rser.2010.11.037

[87] Carballa M, Duran C, Hospido A. Should we pretreat solid waste prior to anaerobic digestion? An assessment of its environmental cost. Environmental Science & Technology. 2011;**45**: 10306-10314. DOI: 10.1021/es201866u

Chapter 6

Techno-Economic Analysis of Biogas Production from Microalgae through Anaerobic Digestion

Na Wu, Cesar M. Moreira, Yingxiu Zhang, Nguyet Doan, Shunchang Yang, Edward J. Phlips, Spyros A. Svoronos and Pratap C. Pullammanappallil

Abstract

Microalgae are a promising feedstock for bioenergy due to higher productivity, flexible growing conditions, and high lipid/polysaccharide content compared to terrestrial biomass. Microalgae can be converted to biogas through anaerobic digestion (AD). AD is a mature technology with a high energy return on energy invested. Microalgae AD can bypass energy intensive dewatering operations that are associated with liquid fuel production from algae. A techno-economic assessment of the commercial feasibility of algae-based biogas production was conducted using Cyanothece BG0011 biomass as an example. BG0011 is a naturally occurring, saline cyanobacterium isolated from Florida Keys. It fixes atmospheric nitrogen and produces exopolysaccharide (EPS). Maximum cell density and EPS concentration of 2.7 and 2.1 g afdw[1]/L (for total algae biomass concentration of 4.8 g afdw/L) were obtained by air sparging. For an areal cell and EPS productivity of 12.4 and 9.6 g afdw/m2/day, respectively, the biomethane production cost was 14.8 $/MMBtu using covered anaerobic lagoon and high-pressure water scrubbing for biogas purification. Electricity production from biogas costs 13 cents/kwh. If areal productivity was increased by 33% from the same system, by sparging air enriched with 1% CO2, then biomethane cost was reduced to 12.16 $/MMBtu and electricity cost to 11 cents/kwh.

Keywords: microalgae, anaerobic digestion, biogas, techno-economic analysis, *Cyanothece* BG0011

1. Introduction

Resource depletion and carbon emissions caused by using fossil fuels have increased interest in alternative fuel sources. Utilization of biomass resources is one option to meet the energy requirements for rapid industrialization and

[1] Ash free dry weight.

population growth with potential environmental and economic benefits. Energy could be derived from a variety of terrestrial, renewable, bio-based feedstocks like sugar-based biomass (e.g. corn, sugarcane, sugarbeet) and lignocellulosic biomass (e.g. wheat straw, corn stover, sugarcane bagasse, forestry residues, switchgrass, energy cane, sorghum, short rotation woody crops). However, production and conversion of these feedstocks could entail risks associated with disruption of the food chain and biodiversity, depletion of freshwater resources and eutrophication.

Aquatic biomass like microalgae is a promising feedstock with many advantages over terrestrial plants. Its use dates to 1940s [1, 2]. To meet an energy shortage during this period, microalgal biomass was proposed to be used as a source for lipids. Microalgae have higher yield from incident solar energy and higher areal productivity. The photosynthetic efficiency of microalgae (around 3–8%) is substantially higher than that of terrestrial plants (typically 0.5%) due to their simple structure and convenient access to nutrients [3–5, 108]. Therefore, less land area is required and non-arable, non-productive land could be used for their cultivation. Some species could be cultivated using low quality water such as seawater, brackish water, desalination reject water and wastewater. A microalgae production facility could be operated as a closed loop system by allowing for recycling of water, nutrients and energy from downstream production processes [6, 7, 144]. Microalgae are characterized by high lipid/starch/protein content with a lack of lignin, which makes them well-suited for different conversion technologies [8–10]. Besides, microalgae cultivation has less potential to interfere with food and feed production. With such versatility, microalgae appear to be a promising biorenewable resource that has the potential to completely replace fossil resources [11]. Research in microalgae biotechnology has increased dramatically since 2005 and has been a very active field in recent years, especially to produce biomass and biofuels [12, 110, 111, 117, 118, 136, 143].

Though microalgae may demonstrate benefits over terrestrial feedstocks, the major challenges for their production include significant utilization of nutrients, high energy input for harvesting and dewatering, and complex downstream conversion processes for usable fuels like ethanol and biodiesel [6, 8, 100, 109, 131]. An alternative which can potentially decrease the energy footprint could be biogas production through anaerobic digestion [122, 125, 127, 137]. Anaerobic digestion (AD) is a biochemical process that mineralizes organic compounds to biogas through the synergistic and concerted action of microorganisms under anaerobic (O_2 free) conditions. Dry biogas is primarily a mixture of methane and carbon dioxide with traces of ammonia, volatile organic compounds and hydrogen sulfide. Methane content of dry biogas usually ranges between 50 and 70% (by volume). Methane has a higher heating value on a mass basis when compared to liquid fuels, such as biodiesel and bioethanol [13, 145]. AD has been recognized as a mature technology to treat organic waste streams and is widely practiced due to its high energy output to input ratio, environmental benefits, as well as for its process simplicity—compared to bioethanol/biodiesel processes [13, 14]. It is suitable for organic feedstock with high moisture content [15] and so can directly be applied to wet algae biomass feedstock with perhaps little dewatering. Besides, no harsh pretreatment is necessary for algal biomass due to the negligible lignin content [14]. The algal biorefinery could be engineered to be resource efficient by recycling phosphorus and nitrogen nutrients in the digestate effluent and carbon dioxide from biogas upgrading processes for microalgae cultivation [13, 14, 16, 17].

In addition to the physical and chemical properties of the fuel as specified by technical standards, the characteristics desired by the stakeholders, distributors and, consumers could also include sustainability indices related to environmental, social and economic performance. Techno-economic analysis (TEA) establishes a capital and operating cost profile to determine the potential economic viability of the production process for realizing its commercial feasibility. It can be an integral tool to direct research during development of specific technology and assist with investment by averting unnecessary expenditures. A number of techno-economic assessments have been completed to evaluate the economic feasibility of biodiesel derived from microalgae [9, 22, 69, 140, 141]. However, there is a lack of techno-economic analysis on anaerobic digestion of microalgae for biogas production, especially full-scale production taking the characteristics of algae species into consideration. In this chapter, the entire production process from algae cultivation to biogas upgrading will be discussed emphasizing the key cost drivers. TEA literature is reviewed for methodology and state of art technologies. An example of TEA was conducted based on the biogas production process from a microalgae/cyanobacteria species *Cyanothece* BG0011 [82].

2. Anaerobic digestion

An anaerobic digestion (AD) process can biochemically convert the whole, wet biomass rather than specific components. The emissions and effluents from the process can be captured for reuse of components like carbon dioxide, ammonia, and phosphorus, and therefore has the potential for economic and environmental benefits. The general biochemical steps in the AD process include: (1) hydrolysis: the breakdown of macromolecules like proteins, lipids, polysaccharides into simpler compounds such as amino acids, sugars, fatty acids and glycerol; (2) acidogenesis and acetogenesis: the hydrolyzed molecules are converted to volatile fatty acids, primarily acetate, hydrogen, and carbon dioxide; (3) methanogenesis: methane production from acetate, hydrogen and carbon dioxide. The hydrolysis step plays a crucial role in determining the successful production of methane [37, 145]. The biochemical processes in AD also occur in nature. AD technology is well established and recognized as a robust technology to convert biomass to bioenergy [146].

Despite the potential, questions related to the economic feasibility and the net energy output are the main hurdles hampering the development of biogas production from microalgae [14, 18–20]. For example, due to the specific structure and composition of the microalgae cell wall, the yield of biogas could be low. Pretreatment to disrupt the cell walls could require high energy inputs. The algae productivity could be low and cultivation cost could be high. Thus, the viability of microalgal biogas production may depend on improvements of efficiency and economic performance. Ongoing efforts include developing inexpensive biomass feedstock, maximizing energy return on investment, and minimizing environmental risks. As only a few studies are available in the literature on the economic feasibility of microalgal biogas exploitation [14], the evaluation and analysis of microalgal biogas production cost will be based on conversion efficiency, technological design aspects as well as available cost information.

3. Key drivers of microalgal biogas production cost

The production of biogas from microalgae feedstock entails a series of steps starting with algae cultivation. Implementation of each step involves capital and operational expenditures. The key drivers such as algal biomass productivities, harvesting and dewatering techniques, AD designs, biogas utilization options, integration of algal production, and AD with other bioprocesses were addressed. The production cost breakdown was illustrated in a harmonized framework and a dynamic connection between the technological and economic/environmental assessments was established.

3.1 Microalgae cultivation, harvesting and dewatering

A photobioreactor is the essential component of an algae cultivation facility. An open raceway pond (ORP) and a closed photobioreactor (PBR) are two major cultivation platforms. These two platforms for algae biomass production have been extensively studied [22–27, 83–85, 101]. The main differences are highlighted in **Table 1**. In addition, the steps from inoculum preparation to obtaining the wet algal paste typically include systems for culture circulation, growth medium supply, air/flue gas supply, culture cooling, culture harvesting, and process monitoring. Heat exchangers, pumps, and a piping network are also required. The location and climate are important factors for algae cultivation.

Due to the high methodological variation of TEA in literature, drawing a generic conclusion over the economic feasibility of microalgal cultivation could be impossible. From the technological and economic perspective, the factors presented below are the ones most prominent in the existing literature and identified as important topics in the development of algae fuels.

1. *Microalgae productivity and culture stability*. According to Davis et al. [23], achievable productivity has a strong influence on the economics. Productivity of more than 25 $g/m^2/day$ annual average is critical for maintaining a relatively low minimum biomass selling price. A significant increase in productivity has to be achieved to reduce cost substantially [25]. The cultured strain should have high growth rate and a steady biomass composition. GMOs or extremophiles could provide culture robustness [22]. However, due to lack of regulations for managing GMOs, it is unlikely permits could be obtained for commercial cultivation of GMO algae strains. For commercial outdoor systems, uncertainties could be associated with seasons of the year and across multiple locations. Thus, the productivity data should be integrated with meteorological data for geographically and seasonally resolved assessments using a robust strain.

2. *Photobioreactor design, construction, and operating conditions*. For the open pond system, pond liners were found to be one of the primary cost contributors [22, 23, 28]. The location of the pond facilities could be selected according to the nature of the soil. For example, ponds built on soil with high native clay content could avoid full liners to reduce the cost. Acién et al. [26] presented a cost analysis of microalgae production using tubular photobioreactors. In these systems, photobioreactors were found to be one of the significant cost contributors. Generally, open raceway ponds are economically advantageous by more than a factor of 2, compared to closed photobioreactors [29]. However, due to increased productivity and culture stability, closed photobioreactors still have the potential for commercial applications.

3. *Energy consumption*. Primary energy consumption is due to the energy required for mixing, circulation, aeration and CO_2 sparging. The energy consumption for mixing at experimental scales usually exceeded commercial-scale

operations requirements and needs to be optimized to determine the minimum energy requirement [27, 28]. Mixing devices such as the paddle wheels are significant capital cost contributors besides the photobioreactors.

4. *Nutrients and carbon dioxide supply*. Higher productivity usually involves higher consumption of nutrients. Thus, nutrient input needs to be adjusted to balance the tradeoff against productivity [23]. Carbon dioxide was found to be the most expensive consumable among the raw materials [26]. Siting algae cultivation facilities on land adjacent to industrial CO_2 sources like flue gases may be effective in reducing cost. However, the substantial logistical and practical constraints of using flue gases in facilities of varying sizes are still a challenge [23].

5. *Land and water*. Even though, microalgae can be cultivated on nonarable land, the soil composition, climate, solar radiation have a substantial influence on their growth. The most suitable location should be warm places or close to the equator where insolation is not less than 3000 h/year [24]. Water is required during algae cultivation to compensate for evaporation or for cooling purposes. Availability of water at low cost is critical for process success. Water reuse, wastewater, seawater, brackish water and reasonable distance to the water source has the potential to reduce costs.

6. *Scaling*. It is critical to quantify the economy of scale for algae production to achieve economic viability [23]. However, large uncertainties and unrealistic assumptions will exist in the research where the productivity potential for microalgae at large-scale is being estimated through linear extrapolation for laboratory-based growth data [30]. Data variability and growth modeling considering geographical information should be considered in large-scale assessments.

7. *Labor and depreciation*. Tredici et al. [21] performed a TEA of the microalga *Tetraselmis suecica* production based on a 1-ha plant in Tuscany, Italy. Cost data were collected from manufacturers and suppliers as well as operating data from pilot and commercial facilities. This study found that the major fraction of cost was labor at small scales (1 ha) and when the pilot plant is scaled to 100 ha, capital expenses contribute the most to the production cost. This assessment is site and strain-specific, but still provides valuable insights for the economic evaluation.

Algae harvesting and dewatering methods include gravity settling, chemical coagulation, flocculation, filtration, centrifuge, and drying. The economic feasibility and energy consumption are two criteria for assessing the performance of unit operations for harvesting and dewatering methods. It was found that the cost of separation takes 20–30% of the biomass production costs [32, 33]. Gravity settling, chemical coagulation, and flocculation usually concentrate the microalgal slurries to 2–7% while filtration and centrifugation concentrate microalgal slurry to 15–25% of total suspended solids [32]. The suitability of microalgal dewatering methods has been investigated for scalability, species flexibility, and downstream processing efficacy [33–36]. Dewatering methods reaching high biomass concentrations are usually associated with high energy input and cost. Thus, a combination of dewatering methods such as flocculation followed by filtration is generally considered to be economical due to the increased harvest efficiency. For downstream processing, methods such as flocculation using flocculants comprised of cationic and anionic

	Open raceway	Closed bioreactor
Biomass productivities	Low	High
Harvesting biomass concentration	Low	High
Total capital cost (CAPEX)	Relatively low	High
Total operational cost (OPEX)	Relatively low	High
Reliability (low contamination risk, stable yield)	Low	High
Net energy ratio (energy output/input)	>1	>1 in some cases
Area required	High	Low
Process control	Low	High
CO_2 loss	High	Low
Water evaporation	High	Low
Photosynthesis efficiency	Low	High
Scale-up	Easy	Difficult

Table 1.
A comparison of the open raceway and closed bioreactor systems for algae cultivation.

poly-electrolytes, synthetic polyacrylamide polymers and starch-based polymers can be employed. However, the detrimental effect of these flocculants on the subsequent microbial processes need to be considered. For example, anaerobic digester stability and gas production could be affected by metal contamination. Future work should include replacing chemical coagulants with natural and low-cost organic ones for harvesting algal biomass.

3.2 Anaerobic digestion systems

3.2.1 Pretreatment

The efficiency of biogas production has been shown to be species-dependent [39]. One crucial factor is the differences in structure of microalgae cell walls. The role of the cell wall in the microbial degradability of algae biomass is highlighted in many investigations [6, 13, 37, 38, 40–43]. Many microalgae species (e.g. *Chlorella kessleri*, *Scenedesmus*) have recalcitrant cell walls, which make it difficult for anaerobic cultures to hydrolyze microalgal intracellular organic matter. Thus, to improve the biodegradability of microalgal biomass, pretreatments methods have been developed to disrupt or solubilize cell walls [112–116]. General insights from these studies are: (1) pretreatment methods are species-specific and their success depends on the nature of the cell wall; (2) mechanical pretreatments which consume electricity are more energy intensive than thermal, chemical and enzyme pretreatments; (3) chemical pretreatments usually have a low cost but produce inhibitory substances which could hamper the AD process; (4) for pretreatment mechanisms such as disruption of microalgal cell walls, the synergistic effects of the enzymes need further investigation; (5) combined pretreatments may provide energy and cost-effective options; (6) multi-objective optimization techniques could be used to obtain a high biogas yield with a positive energy balance; (7) enzyme/biological pretreatments have high selectivity, low inhibitory effects and higher probability of positive energy return [147]; (8) research on pilot/demonstration scale pretreatments is rare; (9) thermal pretreatments have been employed widely and proven to be the most efficient in microalgae pretreatment for AD; and (10) a detailed economic/energy analysis of microalgal AD for biogas production with pretreatment is still missing.

3.2.2 Hydraulic retention time (HRT), organic loading rate (OLR), and reactor configurations

The capital cost of the anaerobic digester could be reduced by using reactors designed for high OLR and low HRT [37]. The OLRs are typically between 1 and 6 g VS/L/d while the HRT varies between 10 and 30 days [37, 38]. Although high OLR will increase the methane productivity, overloading will decrease the biogas production efficiency due to the accumulation of inhibitors such as ammonia and acids [6, 37, 38]. Also, prolonged HRT could lead to ammonia inhibition due to slow liquid removal rate [41], while a low HRT could cause the washout of the anaerobic bacteria community [6]. Thus, an optimized OLR and HRT should be applied to achieve the expected specific methane yield. Possible solutions could be improving anaerobic digester configurations such as using membrane reactors or upflow anaerobic sludge blanket reactors to decouple the OLR and HRT [37, 119] and on-line control of anaerobic digester operation [124]. These have not been applied for digesting algal biomass. Additional costs for land and infrastructure and energy expenditures for heating the digesters should be included in the economic analysis.

3.2.3 Temperature, pH, salinity, sulfur, and lipids content

AD microorganisms can grow in three temperature regimes: (1) psychrophilic (5–20°C); (2) mesophilic (25–45°C); and (3) thermophilic (45–65°C). The temperature effect on AD has been discussed [13, 37, 41]. The beneficial temperature regime for AD operation is anaerobic digester is species-specific [44, 45]. The rate of methane generation can be enhanced under mesophilic and thermophilic conditions. The increased temperature could improve enzymatic activity for degrading microorganisms, and at the same time, the photosynthesis activity of viable microalgae within the digester could be reduced [13, 37]. However, an increase in temperature beyond the tolerable range of each temperature regime could cause inactivation of the microbes. Thermophilic temperature may cause increased hydrolysis of nitrogenous compounds which may increase ammonia levels and in turn can cause inhibition [6]. For large-scale biogas productions, the energy required for heating may be more than 1/3 of the total energy output in the form of biogas [46]. Thus, the net energy production from algae biogas may still be limited due to the high heat input associated with a low concentration of algae substrates.

The pH needs to be maintained at an appropriate level for efficient conversion of biomass to biogas. The growth of microbes, enzyme activity, and the biogas compositions are influenced by the pH [47]. The optimum pH level depends on each step of AD [41]. Generally, the pH values are maintained between 7 and 8 for single stage anaerobic digesters [13, 41].

Microalgae grown in a saline environment offer a sustainable alternative to other biomass by utilizing non-arable land and seawater. Marine microalgae can usually grow in a salinity range of 35–125 ppt [48]. However, when a highly saline culture is processed in an anaerobic digester, the high salinity could be inhibitory to the AD process. The effects of salinity and concentration of sodium are discussed in previous studies [6, 38]. Adaptation of anaerobic digester microbial consortiums under different saline conditions was investigated by Mottet et al. [121]. In a promising study, methane production was observed from anaerobically digesting *Dunaliella salina* biomass at 35 g/L of salinity.

Sulfide is a required micronutrient for anaerobic microorganisms, but high concentrations of sulfide (200 mg/L) could be toxic [6]. For saline microalgal species, the sulfur inhibition may occur due to the presence of oxidized sulfur compounds in saline algae growth medium. Proper inoculum selection for anaerobic digesters could favor the growth of methanogenic bacteria and limit the growth sulfate-reducing bacteria [49].

Lipids can also be inhibitory to the AD process [6, 18, 50] although lipids have a high theoretical methane potential. Generally, inhibition would occur when lipids concentrations are higher than 30%. In this case, the high-lipid microalgae are suitable for lipid extraction for production of liquid fuels.

3.2.4 C/N ratio

Microalgae biomass generally has a higher composition of protein than terrestrial plants [6, 37]. The degradation of protein will cause ammonia accumulation and inhibit the methanogenesis process. The optimum C/N ratio for AD is between 15 and 30 while this C/N ratio for microalgal AD is generally below 10 [13, 38, 41]. Thus, increasing C/N ratio and reducing the ammonia toxicity are important to enhance the biogas yield and productivity from microalgae. Possible solutions to this issue could be; (1) using ammonia-tolerant inoculum generated either by bioaugmentation or by acclimation [37, 38]; (2) using microalgae biomass that was cultivated under nitrogen-limitation [41, 99, 102, 130]; (3) co-digestion with sludge, oil-greases, waste paper and food wastes [13, 41, 54]; and (4) using a two-stage AD for better control of the anaerobic microbial communities [6]. However, these solutions may add more complexity to the system, in which the economic and energetic performance is still clear. For example, the co-substrate needs to be secured for co-digestion; the digester volume and cost may increase due to the loading of the co-substrate; more environmental burdens may be associated with the shipping of biomass, and nitrogen-limitation cultivation may affect microalgae productivity.

3.2.5 Other factors

Many other factors could affect the biogas yield and production of microalgal biomass. For example, the harvesting time influences the composition and biodegradability of algal biomass. Thus, it is essential to harvest algae in the appropriate stage of growth [13]. Storage conditions such as temperature also have an impact on biomass quality like macromolecular distribution and the content of organic compounds. Besides, inoculum to substrate ratio control is instrumental in avoiding inhibition problems such as drop in pH [51].

3.2.6 Biochemical methane potential (BMP) of microalgae biomass

The overall biogas yields depend on the chemical composition of the algae strains. The target strain should be highly digestible. The volatile solids/ash-free dry weight of microalgae plays a significant role in predicting theoretical biogas production potential, which is a critical factor in determining biogas productivities. Theoretically, the methane yield from different components of microalgae is as follows: lipids—1 L CH_4/g VS, proteins—0.85 L CH_4/g VS, carbohydrates—0.42 L CH_4/g VS at standard conditions. Although the lipids have a high theoretical methane yield in AD, a high lipid content (more than 40%) will produce inhibitory substances such as long chain fatty acids [6]. Thus, for high-lipid content microalgae, lipid removal for biofuels production may be a better solution than biomass sent directly to AD.

The impact of the algae cell wall is another critical factor affecting methane yield. Some species either lack cell wall or have cell walls rich in easily-biodegradable proteins as in *Dunaliella salina*, a halophilic microalgae and *Chlamydomonas reinhardtii*, a fresh water green microalgae [38]. Even easily degradable cell wall alone does not ensure efficient methanization. Factors such as the presence of methanogenesis inhibitors, anaerobic microbial community, hydraulic retention

Components	**Species**			
	***Scenedesmus* sp.**	***Chlamydomonas reinhardtii***	***Chlorella vulgaris***	***Nannochloropsis salina***
Protein % of DW	33	65	64	19
Carbohydrates % of DW	35	23	18	45
Lipids % of DW	22	13	10	36

Table 2.
Approximate compositions of four microalgal species: Scenedesmus sp., Chlamydomonas reinhardtii, Chlorella vulgaris, and Nannochloropsis salina [41].

time, organic loading rates, salinity, carbon to nitrogen ratio, and the concentration of digestible substrate will also affect the final methane yield of microalgae.

The microalgal strains which have been investigated extensively include *Scenedesmus*, *Chlamydomonas*, *Chlorella*, and *Nannochloropsis* [12]. The compositions of these four species are shown in **Table 2**. AD conversion process with biochemical methane potential (BMP) to theoretical methane potential (TMP) ratio ≥ 70% are considered highly efficient. *Chlamydomonas reinhardtii* could achieve a 74% BMP (405 ml methane/g VS) to TMP (549 ml methane/g VS) ratio without any pretreatment [52]. Schwede et al. [53] achieved high digestibility of *Nannochloropsis salina* with thermal pretreatment. The methane yield significantly increased from 0.2 to 0.57 m^3 kg VS^{-1} under batch conditions with a BMP to TMP ratio increasing from 31 to 89%. Similarly, *Chlorella vulgaris* shows a significant increase in BMP after pretreatments: from 54 to 85% BMP/TMP ratio [41, 52] under an enzyme pretreatment; and from 62 to 78% BMP/TMP ratio under a biological pretreatment [55, 123]. *Scenedesmus* sp. did not show a BMP/TMP ratio higher than 60%, even after enzyme or thermal pretreatments [56, 57]. The BMP varies from species to species, but no significant difference was found between fresh water microalgae and saline microalgae [58].

4. Techno-economic analysis

In published TEA works, the process complexity was often simplified in terms of limited pathways, few choices of economic drivers and implicit assumptions regarding the growth conditions, process modeling factors and financing of the production facility. Existing reviews in anaerobic digestion of microalgae biomass such as Ward et al. [6] focus on the integration of anaerobic digestion into biodiesel refineries. Considering that diesel or ethanol are more valuable products, anaerobic digestion was suggested to treat the residual biomass to improve the economic viability and sustainability of overall microalgae biodiesel/ethanol stages. Global research in various pathways is going on towards the sustainable development of algae biofuels. The following sections will review these works, highlight the variability of methods of estimating microalgal biogas production cost, find the key drivers of cost contributors, pointing out the convergence and difference in published results, and give a view of the whole value chain towards scaling-up and commercialization when performing a techno-economic analysis (TEA).

4.1 TEA framework

To achieve an optimal facility design, it is necessary to evaluate the tradeoff resulting from the interactions between technical advances and financing parameters. The technical objectives include maximizing microalgal biomass productivity, maximizing biogas yield via AD of biomass, and process stabilization. The economic objectives are to minimize the production cost and maximize the economic benefits. **Figure 1** shows the TEA framework for the sustainability analysis of biogas production from microalgal biomass through anaerobic digestion. The whole biomass processing value chain is determined by the technology framework and progress through experimentally validated process specifics. Economic analysis is based on the process design, which includes the cost assessments and investment analysis. A decision-making platform is built for raw material suppliers, producers and stake holders in an economic perspective. Correspondingly, the economic consequences will direct the research & development of new technologies, which could form a dynamic connection and optimization framework.

Environmental TEA (ETEA) extended the TEA framework with an environmental assessment based on a life cycle analysis [70]. The ETEA is based on the technology readiness level, which means the assessments are performed using the available data based on technology maturity. This would avoid a mismatch between the assessment methodology and the technology readiness level. For example, the whole biogas life cycle includes phases from the biomass cultivation to the final usage and end of life. Under current technology maturity, the whole data set is unavailable, which limits the assessments to certain life cycle phases.

4.2 State of the art: TEA of microalgal biogas

Biorefinery optimization and full utilization of biomass addressing in the economic viability and environmental sustainability of the production of algae biofuels can be found in [39, 71, 72]. Dutta et al. [72] analyzed the sustainability of microalgae-derived biofuel production by performing a TEA and life-cycle assessment and found that coproducts valorization is more energy efficient than the processes focusing on specific components such as lipids. Biorefineries with coproducts and byproducts could have better utilization of the algal biomass and can increase the revenue, thus show greater possibility of achieving economic feasibility. In microalgae biodiesel and bioethanol productions, anaerobic digestion is usually integrated into the biorefinery to treat the residues for energy and nutrient recovery. Sialve et al. [18] compared the energy recovery ratio for two scenarios: direct AD of the whole algae biomass and AD of residue biomass after lipid extraction. Direct AD of

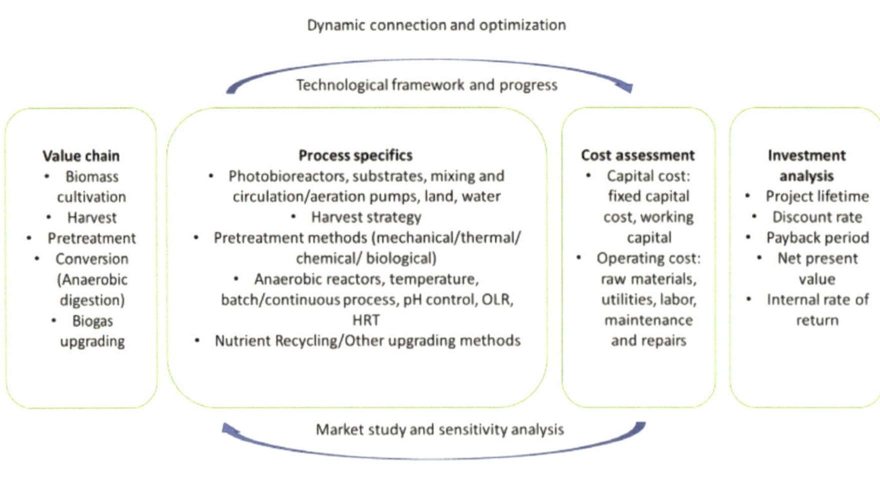

Figure 1.
TEA framework for biogas production from algae biomass.

the whole biomass was considered to have a higher energetic recovery when the cell lipid content does not exceed 40%. Also, increased lipids content in microalgae is not generally compensated with increased productivity due to nitrogen limitation. The potential of direct AD of microalgae biomass was addressed in their research, taking into account the energetic recovery and necessary nutrient recycle for large-scale productions. Chia et al. [73] discussed the economic potential of biohydrogen and biogas production in Germany and Spain. Two processes were compared: direct AD of microalgae biomass (DAD) and coupled hydrogen and biogas production (CHB). In the CHB process, hydrogen was first produced by dark fermentation then effluent from hydrogen fermentation was used for biogas production. The CHB was found to have a lower operating cost due to no additional water and nutrients requirements for the bioreactor feed while the DAD process requires algal biomass in combination with other feedstocks. Both cases have production costs 13–16 times higher than the market price for natural gas. A 1/3 higher biogas yield and a 1/2 lower labor cost did not change the economic status of both processes, due to the high cost of fertilizer and building photobioreactors for microalgae cultivation. Milledge and Heaven [74, 129] performed an energy balance of biogas production from microalgae. Their research emphasized a combination of dewatering methods, as well as the efficient exploitation of the heat generated by the combustion of biogas in combined heat and power (CHP) units to show the energetic viability of the whole process.

Chew et al. [68] assessed the potential of microalgae biorefineries for producing high-value products such as pigments, proteins, lipids, carbohydrates, and vitamins. The high-value products were added to improve the biorefinery economics. Open pond cultivation and medium recycling were mentioned to have better economic performance than other biorefinery structures. Water, land usage and capital cost were challenging for the economic viability of algal biofuels. The high-value products also need to improve aspects such as separations method, energy consumption, and control of product loss. AD was emphasized to recycle a considerable amount of nutrient usage to make microalgal fuels head towards its large-scale production. Several authors [13, 17, 37, 38, 75, 133, 134] synthesized scientific literature on biogas production from algae and suggested integration of the technology with other technologies as well as co-digestion with other substrates for an optimized biorefinery that sustainably produces biogas. Singh and Gu [76] recommended integrated processes that combine algae cultivation and wastewater treatment for methane production, which could offset the higher cost in comparison to methane production from corn and woody biomass.

Zamalloa et al. [8] evaluated the techno-economic potential of methane production from microalgae. The assessment was carried out using high rate anaerobic digesters (10–20 kg COD/m^3/d) and preconcentrated algae biomass from a full-scale open pond. The energy production cost from microalgal biogas was estimated to be 0.087–0.17 euro/kWh with an algae biomass cost of 86–124 euro/tonne. The result was based on a feed-in tariff of 0.133 euro/kWh and a carbon credit of 30 euro/ton of carbon dioxide. This study is one of the limited works that has been done on a comprehensive technological and economic assessment of electrical and thermal energy produced by biogas through AD of microalgae.

Collet et al. [77] performed a life-cycle assessment (LCA) of biogas production from the microalgae *Chlorella vulgaris* and found that electricity consumption and the impacts generated by the production of methane from microalgae are strongly correlated. Decreasing mixing and heating cost in different production steps or increasing the efficiency of AD were important to reduce the overall cost.

The studies surveyed show considerable variability in the calculated fuel cost and identifying the significant cost contributors. The varied results come from different conversion pathways, technical assumptions (productivity, reactor design, process parameters, etc.) and economic factors (interest rate, raw material cost, etc.), diverse environmental and social conditions (consideration of season and location), and validation of sub-process models (lab/pilot plant/commercial scales). Nevertheless, the contributors to the production cost are mainly identified as microalgal strain selection, biomass cultivation and harvesting, AD operating conditions, biogas upgrading methods, waste management, and type of biorefinery. Thomassen et al. [78] evaluated the methodological reason for the wide variation in the results of multiple environmental and economic assessments. They proposed an environmental techno-economic assessment which can help to solve the challenges for a sustainability assessment: framework for methodology, harmonized assumptions, and integration of different dimensions (stages of technological maturity, technological process). This method is based on the dynamic technological process parameters and the same system boundaries for an integrated TEA and LCA.

4.3 Cost management

Gnansounou and Dauriat [79] investigated TEAs following different types of cost management systems in value engineering, target costing and a combination of value engineering and target costing. Value engineering includes process design via data collection and process flowsheeting. Process simulators such as Aspen Plus enables the evaluation of the whole process chain based on scale up of the pilot plant, state of art technologies and price quotes. For microalgae to biogas technologies, key issues along the process chain include the suitable choice and operation options of the microalgae species, harvesting/dewatering strategies, pretreatment methods, AD configurations, recycling the digestate, and energy integration. Not all the steps are necessary for technologies with simplified processes and high economic potential. Target costing is a market-oriented method, which means a target selling price was set for the cost evaluation based on market and societal values. Following the target price, the target cost of the final product and each step of the supply process will be estimated, which means the cost allowance will play a key role in the process design. Target costing could integrate with value engineering in the cost management activities, so the cost allowance and cost target could be reconciliated. In the case of biogas production, the target costing evaluation seems unfeasible for the whole process due to the weak financial position of the natural gas market [80].

Real options analysis framework was employed by Kern et al. [81] for TEA. The model was adapted to accept stochastic price data for energy and agricultural commodities as well as static operating parameters assumptions for the algal biofuel plants. The TEA work was combined with life cycle analysis in a dynamic system—the fluctuations in market prices for energy and agricultural commodities will influence the operation decisions of the biofuels plants and its associated environmental impacts. Areas such as carbon tax, resource shortage and market forces could be investigated for their impact on biofuel plant design and operations in a dynamic system in the future. This gives the stake holders and suppliers more flexibility in making decisions.

4.4 TEA limitations

The limitations of TEA include the potential competition for resources. For example, the microalgae biomass could have non-energy applications and has the potential for producing high value products besides biofuels. Then the biomass cost for the process will be influenced not only by the biomass production activities but also the market price which is determined by both the suppliers and purchase competitors.

The sustainability of biogas production from microalgae will depend on not only the commercial viability but also environmental improvements such as greenhouse gas emission reduction, lack of direct and indirect impacts on land-use as well as biodiversity and eutrophication. The scope of TEA is limited for the environmental impact assessment, while these impact categories are appropriate for the goals of the overall sustainability analysis. Thus, an ETEA would allow assessing the sustainability of the entire value chain. Besides, TEA is not reflecting social impacts such as social awareness of algal biofuels' non-food competitive characteristics, rural development, and public recognition.

5. Case study: TEA of anaerobic digestion of *Cyanothece* BG0011

The microalgae used for this case study is a cyanobacterium, *Cyanothece* sp. BG0011 isolated from a shallow lake in Florida Keys [82]. Compared to other algal species, this species is endowed with unique features. First, cyanobacterium *Cyanothece* sp. BG0011 is a saline species and can be adapted to a wide range of salinities (10–70 psu). Second, it fixes dinitrogen in the air, which means it does not require nitrogenous nutrients in the culture water. Third, it produces exopolysaccharide (EPS) which can be converted to a variety of bioproducts. The aim of this case study is to assess the economic feasibility of biogas production using *Cyanothece* sp. BG0011 as feedstock by conducting a techno-economic assessment. The analysis investigated alternatives to decrease the cost and energy requirement of the cultivation and anaerobic digestion of algae. Utilization of biogas to produce electrical and thermal energy or upgrading to produce pure methane (renewable natural gas) was also considered. A comprehensive TEA was carried out based on experimental data and a set of operational assumptions which could be conceivably achieved in near term. The process flowsheet for biomass to biogas conversion through anaerobic digestion and biogas purification processes was implemented in Aspen Plus V8.8 to obtain mass balance and energy requirement results. The discussion focused on the preliminary exploration of the conceptual design of a microalgae cultivation and bioconversion system as well as an investigation on improvements that could result in the greatest system flexibility, energy yield and cost reductions.

5.1 *Cyanothece* BG0011 cultivation

Results from many experiments [149] conducted in the Bioprocess Engineering Laboratory, Department of Agricultural and Biological Engineering, University of Florida gave an average growth rate of 67.5 mg afdw/L/day (20.25 g afdw/m^2/day) for BG0011 cell biomass and an EPS production rate of 52.5 mg afdw/L/day (15.75 g afdw/m^2/day), resulting in cell density of 2.7 g/L and EPS concentration of 2.1 g/L. The areal rates were calculated by assuming that the depth of culture was 30 cm, which is typically the case for open ponds. In the laboratory, the cultures were cultivated under air sparging, a constant illumination of 1200 μmol photons m^{-2} s^{-1} light and 13 h to 11 h light-dark cycle. Open raceway ponds are generally used for large-scale commercial production of algal biomass [86]. Productivity in industrial-scale raceway ponds is generally lower than in small experimental reactors. In literature, algae biomass productivity performance claims range from 7 to 35 g afdw/m^2/day [23, 87–89] with corresponding net photosynthetic efficiencies from under 1–4%. Among these, for studies involving techno-economic analyses, the baseline productivity assumed was 20 g/m^2/day, with an optimistic value of 25–30 g $afdw/m^2$/day, and a conservative value of 15 g $afdw/m^2$/day. In this study, which assumed that BG0011 is cultivated in current large commercial open ponds, an average productivity of 12.4 g $afdw/m^2$/day (corresponding to a net photosynthetic efficiency of under 1%) was used. Similar growth rates were obtained by [148] when the algae was cultivated by air sparging and exposed to a lower light intensity of 122 μmol photons m^{-2} s^{-1} light and 13 h to 11 h light-dark cycle. Here, laboratory-scale BG0011 cell biomass growth rate is comparable to algae cell growth rates reported from other studies, however, in the case of BG0011, it also produces EPS. The average mass ratio between EPS and cell biomass is 0.778: 1 and also EPS production is cell-growth associated, so for this study it is assumed that in the commercial system, in addition to BG0011 cells, EPS would be concomitantly produced at 0.778 × 12.4 g afdw/m^2/d = 9.6 g $afdw/m^2$/day. The total algae biomass productivity used was 22 g/m^2/d. Henceforth, the term "algae biomass" will include both BG0011 cells and EPS.

The scale of algae cultivation in literature for techno-economic analysis ranges from 200 to 700 ktonne afdw/year [22, 27, 72, 89]. In the present study, the scale of algae cultivation was determined based on a hypothetical 20 million gallons per year ethanol plant. The sugar required for such a plant would be 128 ktonnes afdw/year (assuming yield of around 0.42 g ethanol/g sugar, and 1.1 g sugar/g polysaccharide). Assuming this amount of sugar will be supplied in the form of EPS, the scale of the algae cultivation pond would be 293 ktonnes of algal biomass/year, which also includes BG0011 cell biomass. This scale falls into the range of values found in literature for TEA. To meet a production capacity of 293 ktonnes/year at algal biomass productivity of 22 g $afdw/m^2$/day, land area required would be 3660 hectares (approximately 4 × 4 miles). For a sanity check, this cultivation area was compared to land area required to supplying corn grain for a 20 million gallon per year corn-ethanol plant. Based on annual corn grain yield of 7000 kg/ha with starch content of 72% [150], and assuming a conversion of 0.5 kg ethanol/kg of starch, land required would be 23,700 ha. In this case the total above ground biomass productivity of corn, including corn grain, stover and cobs, is 16,700 kg/ha/year [150] whereas for BG0011 it is anticipated to be 80,300 kg/ha/year.

The BG0011 cultivation cost was estimated based on vendor quotes, literature, or engineering estimates. The installed pond capital cost includes civil work, liner, piping, electrical, other pond costs (such as paddlewheels). In addition, pumps for pumping water from ponds to refinery and for refilling the pond and required land also incur significant capital costs. Plastic lined earthen ponds were chosen for its lower cost compared to concrete ponds. Larger pond sizes would enable economically viable algal biomass production [23]. Here, the installed capital cost was estimated based on "dollars/hectare" of growth ponds for simplicity. The installed pond cost was set to be 80,000 $/ha. Literature value ranges from 46,000 $/ha to more than 150,000 $/ha (value adjusted for inflation) due to different liner scenarios (partial or full) and specific design (e.g. with or without equipment to minimize dead zones) [23, 86] which was not included here. A land cost of 3080 $/acre [90] was used for low-value land. The operation cost for algae cultivation such as utilities, chemicals, labor, overheads, maintenance, insurance tax, etc. were estimated using engineering estimates [91]. BG0011 was assumed to be cultivated in seawater or brackish water. The only fertilizer used for BG 0011 cultivation is phosphorus since it uses dinitrogen in air as a nitrogen source, and seawater would supply rest of micronutrients. From laboratory experiments it was determined that the phosphorous requirement of BG001 is 8.9 mg/L [149], so the annual requirement of phosphorous will be 1186.7 tonnes. Here, triple superphosphate ($Ca(H_2PO_4)_2$ H_2O) which contains 24.6% P is used as phosphorous source with a price of 270 $/tonne (Source: World Bank, 2017). The requirement of triple superphosphate is 4945 tonne/year.

The fixed capital investment was assumed to be borrowed at an interest rate of 10% for 20 years. The plant operates 24 h a day and 360 days annually. The prices were adjusted for Year 2017 using Chemical Engineering Plant Cost Index (CEPCI). These assumptions were also used for the analysis of subsequent biogas production, conversion and upgrading processes. The production cost was calculated as follows:

$$\begin{aligned}&\text{Unit production cost}\\&=(\text{Annual capital charges}+\text{Total operating cost})\\&\quad-\text{Coproduct credits})/\text{Annual production}\end{aligned} \tag{1}$$

Here, the annual capital charges are calculated as follows:

$$\begin{aligned}&\text{Annual capital charges}\\&=\text{Total capital cost}*\text{Interest rate}*(1\\&\quad+\text{Interest rate})\wedge\text{Loan period}/\text{Interest rate}\wedge\text{Loan period}-1\end{aligned} \tag{2}$$

* Total capital cost = Total fixed cost + Working capital.
* Working capital is 10% of fixed capital.

5.2 Anaerobic digestion

The anaerobic digester was designed to treat the un-dewatered whole algae culture from the pond. The energy-intensive steps like algae harvesting and dewatering are avoided in this process which is different from most research [8, 22, 23]. The product biogas was analyzed for economic performance in two different applications: biogas purification or electricity production through combined heat and power.

The first step in modeling mass flow rate of reactor outputs and determining energy requirements is to establish the stoichiometry of reactions. The stoichiometry of methane fermentation of algae biomass was developed based on the following assumptions: (1) microbial cells (cyanobacteria and bacteria) can be represented by the empirical formula $CH_{1.8}O_{0.5}N_{0.2}$ [151]; (2) EPS is pure polysaccharide represented by the empirical formula $C_6H_{10}O_5$; (3) algae biomass can be represented by an empirical formula containing the elements C, H, O and N in the mass ratios in which cells and EPS are produced that is 1:1.2; and (4) methane yield from laboratory assays corresponds to complete decomposition of substrate. The empirical formula for algae biomass was $CH_{1.73}O_{0.67}N_{0.1}$. The stoichiometry for methane formation is written as follows:

$$CH_{1.73}O_{0.67}N_{0.1} + aNH_3 \rightarrow bCH_{1.8}O_{0.5}N_{0.2} + cCH_4 + dCO_2 + eH_2O \tag{3}$$

Methane yield from algae biomass was measured in the laboratory to be 300 ml at STP (g afdw)$^{-1}$. This corresponds to 0.35 moles of methane (mole algae biomass)$^{-1}$, which is equal to value of 'c' in the above stoichiometry. The other stoichiometric coefficients can now be solved from elemental balances for C, H, O and N. The stoichiometry is

$$\begin{aligned}&CH_{1.73}O_{0.67}N_{0.1}+0.17H_2O\rightarrow0.31CH_{1.8}O_{0.5}N_{0.2}\\&+0.35CH_4+0.34CO_2+0.04NH_3\end{aligned} \tag{4}$$

In the anaerobic digester it was assumed that 98% of the algae biomass is converted. Different scenarios (three anaerobic digester types) were investigated to evaluate the economic and energetic performance. A schematic of biorefinery scenarios are shown in **Figure 2**.

Case 1. Above ground mesophilic anaerobic digester. In Aspen, the influent to the reactor was 15 ktonne/h. The temperature was maintained at 37°C. It was operated at an HRT of 25 days.

Case 2. Above ground low-temperature anaerobic digester. Anaerobic digestion at low temperatures (LTAD) was applied to improve the energy balance. In this scenario the digester is operated in the psychrophilic range (12–20°C) [92–94]. However, with the same flow rate, the digester volume is larger to achieve a higher HRT for LTAD than mesophilic and thermophilic anaerobic digestion. Here, the temperature of LTAD is set to 20°C with an HRT of 50 days.

Case 3. Covered anaerobic lagoon. Covered anaerobic lagoon (CAL) does not require additional energy for the biogas production because no heating or mixing processes are involved. Besides, it is economical to construct and operate. The CAL in this research was 6 meters deep and covers an area of 1.5 hectares based on literature data [95]. The HRT was set to 50 days. The cost includes anaerobic lagoon excavation, cut and fill, lagoon liner, inlet and out structures, lagoon cover, ancillaries, pipework & installation, contingencies, design, engineering, etc. Operating costs including utility usage are minimal.

In all three cases above, the capital cost of anaerobic digester was estimated using vendor quotation or literature values. The operating cost was estimated by Aspen Process Economic Analyzer.

5.3 Biogas purification

Several biogas upgrading or purification methods are available such as high-pressure water scrubbing, membrane, pressure swing, gas permeation and chemical scrubbing. High pressure water scrubbing and chemical scrubbing (using amine solutions—MEA) are two of the most commonly used processes.

The MEA scrubbing method uses aqueous monoethanolamine (MEA) for acidic gas removal. The concentration of amine for acidic gas absorption is usually below 30% (by weight). The amine process has two main steps, absorption and stripping [96]. The detailed MEA scrubbing process is shown in **Figure 3**. Raw biogas goes

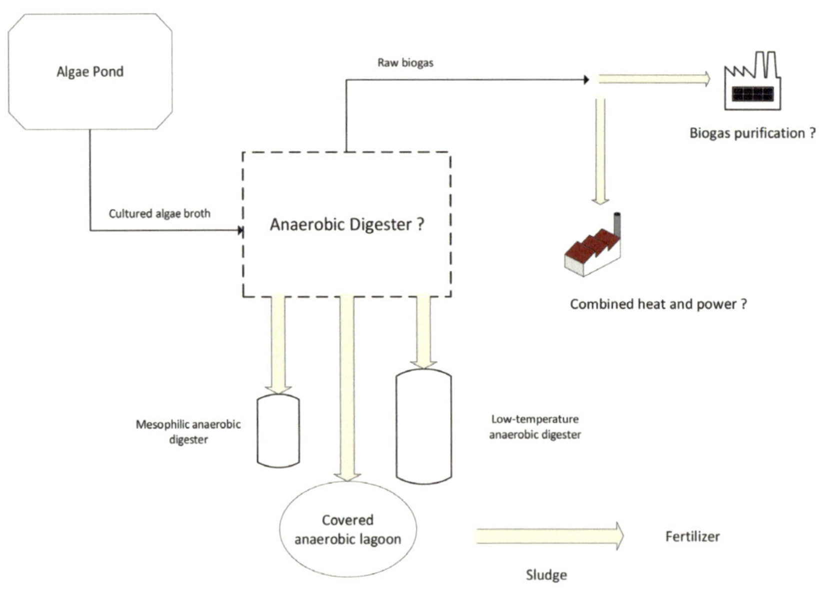

Figure 2.
Schematic diagram showing biorefinery scenarios.

DOI: http://dx.doi.org/10.5772/intechopen.86090

through a scrubbing column in which MEA is flowing counter-current to biogas. The CO_2-rich MEA is collected at the bottom of the scrubbing column and pumped into a stripping column to remove CO_2 and regenerate MEA by heating. Similar to MEA scrubbing, high pressure water scrubbing was also employed for biogas upgrading: biogas is fed to the bottom of scrubber after compressing it to 10 bar. At the top of scrubber, pressurized water is fed. CO_2-rich water is then transferred to a flash column with a lower pressure of 3 bar to release gases for feed recirculation and minimizing methane loss. Then the CO_2-rich water goes through a CO_2 desorption process from the water stream by air [97]. Both biogas purifying approaches

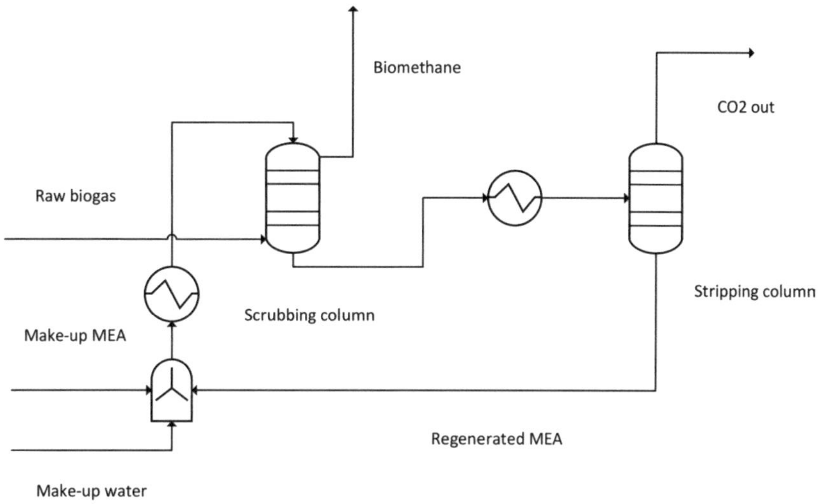

Figure 3.
MEA scrubbing for biogas upgrading.

Specification	MEA	High pressure water scrubbing
Thermodynamic method	ELECNRTL	PSRK
Scrubbing column	RadFrac, 15 stages, pressure: 1.2 bar	RadFrac, 10 stages, pressure: 10 bar
Stripping column	RadFrac, 15 stages, pressure: 8 bar	RadFrac, 10 stages, pressure: 1 bar
Make up chemicals	Water: 150 kmol/h MAE: 750 kmol/h	Water: 11500 kmol/h
Solvent recycle rate	MEA: 0.99	Water: 0.95
Methane loss	1%	0.3%
Product methane purity	95 wt%	99.2 wt%
Capacity (raw biogas flow rate)	948.5 kmol/h	948.5 kmol/h
Capital cost (million $)	8.2	12
Operating cost (million $/year)	20	4.6
Utility cost (million $/year)	17	2
Purification cost ($/kg of methane)	0.3	0.09

Table 3.
Technical and economic aspects of the biogas purifying systems in ASPEN V 8.8.

were simulated in ASPEN Plus to determine the economics of each approach. The technical specification details are shown in **Table 3**. The table shows high pressure water scrubbing to be a more economical alternative and was chosen for the integrated process.

5.4 Power generation from biogas

While the raw biogas can be purified to obtain biomethane, another option is to use the raw biogas to produce heat and power. Steam and electricity can be generated by burning the raw biogas through a combined heat and power (CHP) system. For reference, the CHP system uses General Electric Jenbacher JGS 420 system which is a 1425 kw generator. The total capital cost is $ 1,150,000 (including installation, tax, etc. 2007), which is 807 $/kw. The working capital is 10% of the total capital. The operating cost includes direct operating cost such as operating labor, supervised labor, maintenance and repairs, as well as indirect operating cost such as overhead, taxed, insurances. It is assumed that 40% biogas energy is for electricity, 50% for steam, 10% loss.

5.5 Techno-economic analysis of integrated system

5.5.1 Biomass cultivation economics

The BG0011 cultivation economics analysis details are shown in **Table 4**. In the literature algae production costs range from 150 to 6000 $/tonne [19, 22, 27, 72, 89, 142], however, the studies vary from assumptions (production scale, chemical prices, plant life, etc.) to differences in technical specification (photobioreactor design, algal species, etc.). Some of the estimates also account costs for dewatering of algae [22, 27]. Thus, it is difficult to make a direct comparison between different studies. Besides, specific assumptions in each study could be based on different social-economic conditions, which makes comparisons more complicated [98].

Parameters	**Values**
Production scale	
BG0011 cells production (ktonne/year)	165
BG0011 EPS production (ktonne/year)	128
Total algae biomass production (ktonne/year)	**293**
Capital cost (including fixed, installed and working capital)	
Pond (million $)	308
Land (million $)	26.6
Pump (million $)	7.85
Total capital cost (million $)	342.45
Annual capital charges (million $/year)	**40.22**
Operating cost	
Chemicals (P fertilizer: Ca $(H_2PO_4)_2$ H_2O) (million $/year)	1.3
Other operating cost (including utilities, maintenance and repairs, labor etc.) (million $/year)	3.26
Total operating cost (million $/year)	**4.56**
BG0011 algae biomass production cost ($/tonne)	153

Table 4.
Algae cultivation economics.

5.5.2 Economics of anaerobic digestion

Details of the production cost of renewable natural gas for the three anaerobic digestion scenarios are shown in **Table 5**. Case 2 contains two scenarios: The size of anaerobic digester in Case 2(a) is two times of that in Case 1. This is because the hydraulic retention time is longer under lower temperature, the volume of digester needs to be larger to keep the same production scale (the inflow rate). The size of anaerobic digester in Case 2(b) is the same as Case 1. Keeping the digester volume same as Case 1, because the temperature is lower, the productivity will be lower as well. Thus Case 2(b) has a lower production scale compared to other cases. The effect of temperature was incorporated by using the empirical relationship that for every 10°C rise in temperature the degradation rate is doubled. As the difference between the temperature for Case 1 and Case 2 is 17°C, it is expected that in Case 1, the digester has a processing capacity twice as much as that of the digester in Case 2b. The main contributor to the production cost of biogas is the biomass cost. Considering a carbon credit of 10 $/tonne of CO_2, the production cost of biogas only drops 0.5 $/MMBtu. The results are comparable to Zamalloa et al.'s [8] research (the only paper focusing on the economics of renewable energy through AD, to our best knowledge): 32.2–61.5 $/MMBtu with the algae biomass cost of 115.4–166.4 $/tonne (0.087–0.17 euro/kwh with an algae biomass cost of 86–124 euro/tonne, 2011). The methane yield is 0.012 MMBtu/kg of VS biomass, which is in close agreement to our experimental result 0.0124 MMBtu/kg of VS biomass.

Item	Case 1 (mesophilic anaerobic digester)	Case 2(a) (low-temperature anaerobic digester)	Case 2(b) (low-temperature anaerobic digester	Case 3 (covered anaerobic lagoon)
Biogas production scale (10^6 MMBtu/year)	3.7	3.7	1.85	3.7
Fixed capital cost of anaerobic digester (million $)	67.12	102	67.12	7.5
Capital cost except anaerobic digester (million $)	16.3	16.3	12.3	16.4 (including land: $11400)
Annual capital charges (million $/year)	9.8	13.9	9.3	2.8
Total raw materials (algae biomass) cost (million $/year)	44.8	44.8	44.8	44.8
Other operating (labor, utility, indirect, etc.) cost (million $/year)	25.8	7.1	4.4	7.1
Utility cost (million $/year)	21	2.3	1.4	2.3
Renewable natural gas production cost ($/MMBtu)	21.7	17.8	31.6	14.8

Table 5.
Process and economic assessment for purified biogas production through anaerobic digestion of Cyanothece BG0011 biomass.

Item	Value
Electricity capacity (million kwh/year)	435
Total capital cost of the CHP system (million $) (including fix capital cost and 10% working capital)	52.4
Capital charges (million $/year)	6.2
Steam credits (million $/year)	3.7
Raw biogas cost (million $/year)	47.7
Other operating cost (million $/year)	9.5
Electricity production cost ($/kwh)	0.13

Table 6.
The economics of biogas—electricity and steam system.

5.5.3 Electricity production cost

On an energy potential basis, 40% of total methane produced per year could support a 50 MW power plant. Current residential electricity price is around 12 cents/kwh, while industrial price is around 7 cents/kwh. As shown in **Table 6**, the electricity production cost from biogas is 13 cents/kwh. Renewable energy technologies are usually more expensive than fossil fuel technologies. The reasons could be environmental costs associated with fossil fuels that are not paid by the rate payers, mechanical difficulty in bioenergy production, start-up issues and so on. European countries such as Germany and UK governments subsidize the production of renewable energy by introducing feed-in tariffs. These tariffs may be important to make bioenergy industry profitable.

6. Cost minimization approaches

6.1 Nutrient recycling and biogas upgrading

Nutrient (mostly nitrogen and phosphorous) recycling such as utilizing the digestate or wastewater for microalgae cultivation was highlighted in various studies [59–63, 104–107, 126, 128, 138, 139]. Recycling the effluent from the anaerobic digester for algae cultivation could mitigate the costs associated with supplying nutrient for algal biomass growth and effluent treatment. Erkelens et al. [59] validated that microalgae *Tetraselmis* sp. could utilize its digested effluent as a growth medium and thus form a closed loop system. Also, Prajapati et al. [60] showed that algal liquid digestate have good potential to be utilized as nutrient supplement (30% concentration) in rural sector wastewater for biomass cultivation. The biomass production level is closer to the case in which conventional medium is used. Although there are still technological obstacles when growing microalgae on digestate such as low growth rate due to poor nutrient ratios, shading, ammonia inhibition and bacteria growth, the performance of the nutrient recycling process could be further developed by scale up/optimizing strategies such as controlling inoculum and substrate concentrations, bacteria growth as well as harvesting strategies [59, 61, 64, 132].

One option to increase algae biomass productivity and its concentration in the culture is to enrich the air with CO_2. It has been shown that enriching the air with 1% CO_2 increases cell concentration to 3.46 g afdw/L and EPS concentration to 2.91 g afdw/L, giving an algae biomass concentration of at least 6.37 g afdw/L

[149], which is 1.33 times more than that used in the case study above. The increased productivity of algae biomass will reduce further the cost for biomass production. The CO_2 released from the biogas upgrading process or waste gases from biogas combustion containing CO_2 could be recycled to the algae growth ponds for enriching the air. The economic analysis for this scenario was also performed assuming algae biomass concentration is 1.33 times the previous value of 293 ktonne/year. The estimated production cost for *Cyanothece BG 0011* algae biomass is 121.6 $/tonne. This was calculated by accounting for the following additional costs: (1) capital cost associated with pipes and pumps to take CO_2 from biogas purification system or biogas combustion output to the pond and (2) operating costs resulting from more nutrient addition to maintain higher cell density and power consumption of compressing CO_2 for sparging [152]. Only biogas production from covered anaerobic lagoon as in Case 3 was considered here. Algae production cost was lowered by 20%. The estimated cost of renewable natural gas is now reduced to 12.16 $/MMBtu and the electricity production cost from biogas is only 10.98 cents/kwh.

Upgrading biogas by fixation of the CO_2 in biogas via photosynthesis by microalgae has been investigated with respect to CO_2 removal capability, biomass productivity and O_2 desorption minimization [16, 63–67]. Toledo-Cervantes et al. [16] optimized the biogas upgrading process by studying the influence of the recycling liquid to biogas ratio. The biomethane produced met specification for injection into natural gas grids. However, this technique requires closed photobioreactors. Hydrogen sulfide (H_2S) is another contaminant to be removed from the biogas. Hydrogen sulfide removal was realized by the oxidation of H_2S to sulfate by sulfur oxidizing bacteria that used the oxygen produced photosynthetically in situ. In this case, the algae-bacteria symbiosis was employed in the photobioreactors [67]. Nutrient recycling and biogas upgrading provides not only the opportunity for AD of microalgal biomass to be cost-effective, but also the potential to reduce the environmental impacts.

To move industrial application of biogas production from microalgal biomass towards commercialization, additional assessment is required regarding large scale operations. These include (1) strain robustness, outdoor productivity, location and seasonal effects, yield from real production systems, and harvesting strategy for algae cultivation (2) for biomass to biogas conversion processes, the conceptual process design needs to take the following factors into consideration: costs associated with digester heating, land, and infrastructure as well as operational parameters such as maintaining pH, temperature, mixing, power consumption, and production of coproducts like fertilizer.

6.2 Dynamic growth models

The uncertainty of large-scale algae cultivation is still a challenge which prevents commercialization; process modeling could provide useful information about the performance of microalgae cultivation systems by estimation and optimization of microalgae productivity under different conditions [103]. A growth kinetic model is critical in a process model simulating microalgae cultivation which has a direct impact on downstream conversion processing systems [135] Lee et al. [31] classified the existing kinetic models into three groups: a single limiting substrate (phosphorus, or dissolved CO_2 concentration), a physical limiting factor (light intensity or temperature), and multiple factors (e.g. both substrate and light). Based on their study, there was a tradeoff between the accuracy of the model representation and real-world usability. A future modeling framework should consider along with limiting nutrients, integration of light and temperature, and incorporation of species diversity.

6.3 Biorefinery concepts

AD can be integrated to biorefineries which produce high value products from algae such as chemicals for cosmetics, nutraceuticals and pharmaceuticals. This requires diversified business strategies which benchmark the market potential for the total raw materials and alternative products. In the economic perspective, three approaches could be possible for the development of microalgae AD: (1) implementing AD for biogasification of cell debris or waste streams in microalgal based processes such as biodiesel/bioethanol/high-value bioproducts (e.g. PHA)/fuel cell/hydrothermal liquefaction/hydrogen production [68, 120]; (2) investigation of high-value products from intermediate metabolites produced during AD such as carboxylic acids [37]; (3) electricity production from microalgae derived biogas. In previous sections, the cost of electricity from microalgae derived biogas is comparable with market value while cost of the renewable natural gas from microalgae is much higher than the current market value of natural gas.

7. Conclusion and future work

This chapter reviewed the literature on TEA of biogas production from algae. The key drivers to the overall production cost were identified and possible process improvements to reduce cost were discussed. The need for harmonization of resource, life cycle and techno-economic assessments in the methodology of TEA was highlighted. Modeling efforts, based on well-informed, rigorous engineering-based process models, should be integrated on a baseline framework such that different process technologies, subprocesses and alternative pathways can be directly compared at a system level. TEA model improvements include strategic planning and using reliable input data from simple mass balance calculations to geographically and seasonally specific assessments, as well as risk analysis for large-scale productivity. Nutrient recycling process has the potential to reduce both cost and environmental burdens.

The cultivation of microalgae BG0011 and its economic feasibility as an energy source through anaerobic digestion was evaluated through a techno-economic analysis. The main contribution to the biogas cost is the biomass production cost. The best-case estimate was a biomethane production cost of 14.8 $/MMBtu using covered anaerobic lagoon and high-pressure water scrubbing purification. The cost of electricity production from biogas was estimated to be 13 cents/kwh. Even though these costs are higher than commercial prices in the United States, these are much lower than those costs with production of liquid fuels like ethanol or biodiesel from algae.

Improved algal biomass productivities could be essential for lowering the cost of algae-derived biogas. This could be achieved by recycling the CO_2 released during biogas upgrading or combustion for algae cultivation. Algal biogas economics could be further improved by marketing the digester sludge as a soil-amendment product, considering that nitrogen in the sludge was fixed from atmospheric dinitrogen.

Acknowledgements

The authors gratefully acknowledge funding provided by Office of Energy, Florida Department of Agricultural and Consumer Services under contract number 92420 for this project.

Conflict of interest

The authors declare that there are no potential financial or other interests that could be perceived to influence the outcomes of the research.

Author details

Na Wu[1*], Cesar M. Moreira[1,2], Yingxiu Zhang[1,3], Nguyet Doan[1,4], Shunchang Yang[1], Edward J. Phlips[5], Spyros A. Svoronos[6] and Pratap C. Pullammanappallil[1]

1 Department of Agricultural and Biological Engineering, University of Florida, Gainesville, FL, USA

2 Faculty of Mechanical Engineering and Production Sciences, ESPOL Polytechnic University, Guayaquil, Ecuador

3 School of Environmental Science and Engineering, Tianjin University/ China-Australia Centre for Sustainable Urban Development, Tianjin, China

4 An Giang University, An Giang, Vietnam

5 Department of Fisheries and Aquatic Science, University of Florida, Gainesville, FL, USA

6 Department of Chemical Engineering, University of Florida, Gainesville, FL, USA

*Address all correspondence to: wuna8703@ufl.edu

References

[1] Wijffels RH, Barbosa MJ. An outlook on microalgal biofuels. Science. 2010;**329**(5993):796-799

[2] Borowitzka MA. High-value products from microalgae—Their development and commercialisation. Journal of Applied Phycology. 2013;**25**(3):743-756

[3] Brennan L, Owende P. Biofuels from microalgae—A review of technologies for production, processing, and extractions of biofuels and co-products. Renewable and Sustainable Energy Reviews. 2010;**14**(2):557-577

[4] Su Y, Song K, Zhang P, Su Y, Cheng J, Chen X. Progress of microalgae biofuel's commercialization. Renewable and Sustainable Energy Reviews. 2017;**74**:402-411

[5] Rahimpour MR, Biniaz P, Makarem MA. Integration of microalgae into an existing biofuel industry. In: Bioenergy Systems for the Future. Sawston, Cambridge: Woodhead Publishing. 2017. pp. 481-519

[6] Ward AJ, Lewis DM, Green FB. Anaerobic digestion of algae biomass: A review. Algal Research. 2014;**5**:204-214

[7] Hise AM, Characklis GW, Kern J, Gerlach R, Viamajala S, Gardner RD, et al. Evaluating the relative impacts of operational and financial factors on the competitiveness of an algal biofuel production facility. Bioresource Technology. 2016;**220**:271-281

[8] Zamalloa C, Vulsteke E, Albrecht J, Verstraete W. The techno-economic potential of renewable energy through the anaerobic digestion of microalgae. Bioresource Technology. 2011;**102**(2):1149-1158

[9] Silva C, Soliman E, Cameron G, Fabiano LA, Seider WD, Dunlop EH, et al. Commercial-scale biodiesel production from algae. Industrial & Engineering Chemistry Research. 2013;**53**(13):5311-5324

[10] Moreno-Garcia L, Adjallé K, Barnabé S, Raghavan GS. Microalgae biomass production for a biorefinery system: Recent advances and the way towards sustainability. Renewable and Sustainable Energy Reviews. 2017;**76**:493-506

[11] Milano J, Ong HC, Masjuki HH, Chong WT, Lam MK, Loh PK, et al. Microalgae biofuels as an alternative to fossil fuel for power generation. Renewable and Sustainable Energy Reviews. 2016;**58**:180-197

[12] Garrido-Cardenas JA, Manzano-Agugliaro F, Acien-Fernandez FG, Molina-Grima E. Microalgae research worldwide. Algal Research. 2018;**35**:50-60

[13] Jankowska E, Sahu AK, Oleskowicz-Popiel P. Biogas from microalgae: Review on microalgae's cultivation, harvesting and pretreatment for anaerobic digestion. Renewable and Sustainable Energy Reviews. 2017;**75**:692-709

[14] Montingelli ME, Tedesco S, Olabi AG. Biogas production from algal biomass: A review. Renewable and Sustainable Energy Reviews. 2015;**43**:961-972

[15] Raheem A, Prinsen P, Vuppaladadiyam AK, Zhao M, Luque R. A review on sustainable microalgae based biofuel and bioenergy production: Recent developments. Journal of Cleaner Production. 2018;**181**:42-59

[16] Toledo-Cervantes A, Serejo ML, Blanco S, Pérez R, Lebrero R, Muñoz R. Photosynthetic biogas upgrading to bio-methane: Boosting nutrient recovery via biomass productivity control. Algal Research. 2016;**17**:46-52

[17] Dębowski M, Zieliński M, Grala A, Dudek M. Algae biomass as an alternative substrate in biogas production technologies. Renewable and Sustainable Energy Reviews. 2013;**27**:596-604

[18] Sialve B, Bernet N, Bernard O. Anaerobic digestion of microalgae as a necessary step to make microalgal biodiesel sustainable. Biotechnology Advances. 2009;**27**(4):409-416

[19] Ribeiro LA, da Silva PP, Mata TM, Martins AA. Prospects of using microalgae for biofuels production: Results of a Delphi study. Renewable Energy. 2015;**75**:799-804

[20] Suganya T, Varman M, Masjuki HH, Renganathan S. Macroalgae and microalgae as a potential source for commercial applications along with biofuels production: A biorefinery approach. Renewable and Sustainable Energy Reviews. 2016;**55**:909-941

[21] Tredici MR, Rodolfi L, Biondi N, Bassi N, Sampietro G. Techno-economic analysis of microalgal biomass production in a 1-ha Green Wall Panel (GWP®) plant. Algal Research. 2016;**19**:253-263

[22] Hoffman J, Pate RC, Drennen T, Quinn JC. Techno-economic assessment of open microalgae production systems. Algal Research. 2017;**23**:51-57

[23] Davis R, Markham J, Kinchin C, Grundl N, Tan EC, Humbird D. Process Design and Economics for the Production of Algal Biomass: Algal Biomass Production in Open Pond Systems and Processing Through Dewatering for Downstream Conversion. Golden, CO (United States): National Renewable Energy Lab (NREL); 2016

[24] Slade R, Bauen A. Micro-algae cultivation for biofuels: Cost, energy balance, environmental impacts and future prospects. Biomass and Bioenergy. 2013;**53**:29-38

[25] Richardson JW, Johnson MD, Zhang X, Zemke P, Chen W, Hu Q. A financial assessment of two alternative cultivation systems and their contributions to algae biofuel economic viability. Algal Research. 2014;**4**:96-104

[26] Acién FG, Fernández JM, Magán JJ, Molina E. Production cost of a real microalgae production plant and strategies to reduce it. Biotechnology Advances. 2012;**30**(6):1344-1353

[27] Norsker NH, Barbosa MJ, Vermuë MH, Wijffels RH. Microalgal production—A close look at the economics. Biotechnology Advances. 2011;**29**(1):24-27

[28] Rogers JN, Rosenberg JN, Guzman BJ, Oh VH, Mimbela LE, Ghassemi A, et al. A critical analysis of paddlewheel-driven raceway ponds for algal biofuel production at commercial scales. Algal Research. 2014;**4**:76-88

[29] Quinn JC, Davis R. The potentials and challenges of algae based biofuels: A review of the techno-economic, life cycle, and resource assessment modeling. Bioresource Technology. 2015;**184**:444-452

[30] Moody JW, McGinty CM, Quinn JC. Global evaluation of biofuel potential from microalgae. Proceedings of the National Academy of Sciences. 2014;**21**:201321652

[31] Lee E, Jalalizadeh M, Zhang Q. Growth kinetic models for microalgae cultivation: A review. Algal Research. 2015;**12**:497-512

[32] Barros AI, Gonçalves AL, Simões M, Pires JC. Harvesting techniques applied to microalgae: A review. Renewable and Sustainable Energy Reviews. 2015;**41**:1489-1500

[33] Laamanen CA, Ross GM, Scott JA. Flotation harvesting of microalgae. Renewable and Sustainable Energy Reviews. 2016;**58**:75-86

[34] Fasaei F, Bitter JH, Slegers PM, van Boxtel AJ. Techno-economic evaluation of microalgae harvesting and dewatering systems. Algal Research. 2018;**31**:347-362

[35] Singh G, Patidar SK. Microalgae harvesting techniques: A review. Journal of Environmental Management. 2018;**217**:499-508

[36] Gerardo ML, Van Den Hende S, Vervaeren H, Coward T, Skill SC. Harvesting of microalgae within a biorefinery approach: A review of the developments and case studies from pilot-plants. Algal Research. 2015;**1**(11):248-262

[37] Gonzalez-Fernandez C, Sialve B, Molinuevo-Salces B. Anaerobic digestion of microalgal biomass: Challenges, opportunities and research needs. Bioresource Technology. 2015;**198**:896-906

[38] Wirth R, Lakatos G, Böjti T, Maróti G, Bagi Z, Rákhely G, et al. Anaerobic gaseous biofuel production using microalgal biomass—A review. Anaerobe. 2018;**52**:1-8

[39] Jones CS, Mayfield SP. Algae biofuels: Versatility for the future of bioenergy. Current Opinion in Biotechnology. 2012;**23**(3):346-351

[40] Passos F, Carretero J, Ferrer I. Comparing pretreatment methods for improving microalgae anaerobic digestion: Thermal, hydrothermal, microwave and ultrasound. Chemical Engineering Journal. 2015;**279**:667-672

[41] Klassen V, Blifernez-Klassen O, Wobbe L, Schlueter A, Kruse O, Mussgnug JH. Efficiency and biotechnological aspects of biogas production from microalgal substrates. Journal of Biotechnology. 2016;**234**:7-26

[42] Kendir E, Ugurlu A. A comprehensive review on pretreatment of microalgae for biogas production. International Journal of Energy Research. 2018;**42**:3711-3731

[43] Córdova O, Santis J, Ruiz-Fillipi G, Zuñiga ME, Chamy R, Fermoso FG. Microalgae digestive pretreatment for increasing biogas production. Renewable and Sustainable Energy Reviews. 2018;**82**:2806-2813

[44] Kinnunen HV, Koskinen PE, Rintala J. Mesophilic and thermophilic anaerobic laboratory-scale digestion of Nannochloropsis microalga residues. Bioresource Technology. 2014;**155**:314-322

[45] Zamalloa C, Boon N, Verstraete W. Anaerobic digestibility of *Scenedesmus obliquus* and *Phaeodactylum tricornutum* under mesophilic and thermophilic conditions. Applied Energy. 2012;**92**:733-738

[46] Carlini M, Mosconi EM, Castellucci S, Villarini M, Colantoni A. An economical evaluation of anaerobic digestion plants fed with organic agro-industrial waste. Energies. 2017;**10**(8):1165

[47] Saratale RG, Kumar G, Banu R, Xia A, Periyasamy S, Saratale GD. A critical review on anaerobic digestion of microalgae and macroalgae and co-digestion of biomass for enhanced methane generation. Bioresource Technology. 2018;**9**:319-332

[48] Ishika T, Moheimani NR, Bahri PA. Sustainable saline microalgae co-cultivation for biofuel production: A critical review. Renewable and Sustainable Energy Reviews. 2017;**78**:356-368

[49] Aspe E, Marti MC, Roeckel M. Anaerobic treatment of fishery

wastewater using a marine sediment inoculum. Water Research. 1997;**31**(9):2147-2160

[50] Cirne DG, Paloumet X, Björnsson L, Alves MM, Mattiasson B. Anaerobic digestion of lipid-rich waste—Effects of lipid concentration. Renewable Energy. 2007;**32**(6):965-975

[51] Zhao B, Ma J, Zhao Q, Laurens L, Jarvis E, Chen S, et al. Efficient anaerobic digestion of whole microalgae and lipid-extracted microalgae residues for methane energy production. Bioresource Technology. 2014;**161**:423-430

[52] Mahdy A, Mendez L, Ballesteros M, González-Fernández C. Enhanced methane production of *Chlorella vulgaris* and *Chlamydomonas reinhardtii* by hydrolytic enzymes addition. Energy Conversion and Management. 2014;**85**:551-557

[53] Schwede S, Rehman ZU, Gerber M, Theiss C, Span R. Effects of thermal pretreatment on anaerobic digestion of *Nannochloropsis salina* biomass. Bioresource Technology. 2013;**143**:505-511

[54] Lu D, Zhang XJ. Biogas production from anaerobic codigestion of microalgae and septic sludge. Journal of Environmental Engineering. 2016;**142**(10):04016049

[55] He S, Fan X, Katukuri NR, Yuan X, Wang F, Guo RB. Enhanced methane production from microalgal biomass by anaerobic bio-pretreatment. Bioresource Technology. 2016;**204**:145-151

[56] Mahdy A, Mendez L, Tomás-Pejó E, del Mar Morales M, Ballesteros M, González-Fernández C. Influence of enzymatic hydrolysis on the biochemical methane potential of *Chlorella vulgaris* and Scenedesmus sp. Journal of Chemical Technology & Biotechnology. 2016;**91**(5):1299-1305

[57] González-Fernández C, Sialve B, Bernet N, Steyer JP. Thermal pretreatment to improve methane production of Scenedesmus biomass. Biomass and Bioenergy. 2012;**40**:105-111

[58] Frigon JC, Matteau-Lebrun F, Abdou RH, McGinn PJ, O'Leary SJ, Guiot SR. Screening microalgae strains for their productivity in methane following anaerobic digestion. Applied Energy. 2013;**108**:100-107

[59] Erkelens M, Ward AJ, Ball AS, Lewis DM. Microalgae digestate effluent as a growth medium for Tetraselmis sp. in the production of biofuels. Bioresource Technology. 2014;**167**:81-86

[60] Prajapati SK, Kumar P, Malik A, Vijay VK. Bioconversion of algae to methane and subsequent utilization of digestate for algae cultivation: A closed loop bioenergy generation process. Bioresource Technology. 2014;**158**:174-180

[61] Cai T, Park SY, Racharaks R, Li Y. Cultivation of *Nannochloropsis salina* using anaerobic digestion effluent as a nutrient source for biofuel production. Applied Energy. 2013;**108**:486-492

[62] García D, Posadas E, Grajeda C, Blanco S, Martínez-Páramo S, Acién G, et al. Comparative evaluation of piggery wastewater treatment in algal-bacterial photobioreactors under indoor and outdoor conditions. Bioresource Technology. 2017;**245**:483-490

[63] Prandini JM, da Silva ML, Mezzari MP, Pirolli M, Michelon W, Soares HM. Enhancement of nutrient removal from swine wastewater digestate coupled to biogas purification by microalgae Scenedesmus spp. Bioresource Technology. 2016;**202**:67-75

[64] Uggetti E, Sialve B, Latrille E, Steyer JP. Anaerobic digestate as substrate for microalgae culture: The role of ammonium concentration on the

microalgae productivity. Bioresource Technology. 2014;**152**:437-443

[65] Converti A, Oliveira RP, Torres BR, Lodi A, Zilli M. Biogas production and valorization by means of a two-step biological process. Bioresource Technology. 2009;**100**(23):5771-5776

[66] Posadas E, Serejo ML, Blanco S, Pérez R, García-Encina PA, Muñoz R. Minimization of biomethane oxygen concentration during biogas upgrading in algal–bacterial photobioreactors. Algal Research. 2015;**12**:221-229

[67] Bahr M, Díaz I, Dominguez A, González Sánchez A, Muñoz R. Microalgal-biotechnology as a platform for an integral biogas upgrading and nutrient removal from anaerobic effluents. Environmental Science & Technology. 2013;**48**(1):573-581

[68] Chew KW, Yap JY, Show PL, Suan NH, Juan JC, Ling TC, et al. Microalgae biorefinery: High value products perspectives. Bioresource Technology. 2017;**229**:53-62

[69] Manganaro JL, Lawal A, Goodall B. Techno-economics of microalgae production and conversion to refinery-ready oil with co-product credits. Biofuels, Bioproducts and Biorefining. 2015;**9**(6):760-777

[70] Thomassen G, Vila UE, Van Dael M, Lemmens B, Van Passel S. A techno-economic assessment of an algal-based biorefinery. Clean Technologies and Environmental Policy. 2016;**18**(6):1849-1862

[71] Dong T, Knoshaug EP, Davis R, Laurens LM, Van Wychen S, Pienkos PT, et al. Combined algal processing: A novel integrated biorefinery process to produce algal biofuels and bioproducts. Algal Research. 2016;**19**:316-323

[72] Dutta S, Neto F, Coelho MC. Microalgae biofuels: A comparative study on techno-economic analysis & life-cycle assessment. Algal Research. 2016;**20**:44-52

[73] Chia SR, Chew KW, Show PL, Yap YJ, Ong HC, Ling TC, et al. Analysis of economic and environmental aspects of microalgae biorefinery for biofuels production: A review. Biotechnology Journal. 2018;**13**:1700618

[74] Milledge JJ, Heaven S. Energy balance of biogas production from microalgae: Effect of harvesting method, multiple raceways, scale of plant and combined heat and power generation. Journal of Marine Science and Engineering. 2017;**5**(1):9

[75] Murphy JD, Drosg B, Allen E, Jerney J, Xia A, Herrmann C. A perspective on algal biogas. IEA Bioenergy. [Internet]. 2015. Available from: http://task37.ieabioenergy.com/files/daten-redaktion/download/Technical%20Brochures/AD_of_Algae_ebook_end.pdf [Accessed: 2018-12-05]

[76] Singh J, Gu S. Commercialization potential of microalgae for biofuels production. Renewable and Sustainable Energy Reviews. 2010;**14**(9):2596-2610

[77] Collet P, Hélias A, Lardon L, Ras M, Goy RA, Steyer JP. Life-cycle assessment of microalgae culture coupled to biogas production. Bioresource Technology. 2011;**102**(1):207-214

[78] Thomassen G, Van Dael M, Lemmens B, Van Passel S. A review of the sustainability of algal-based biorefineries: Towards an integrated assessment framework. Renewable and Sustainable Energy Reviews. 2017;**68**:876-887

[79] Gnansounou E, Dauriat A. Technoeconomic analysis of lignocellulosic ethanol. InBiofuels. Cambridge, Massachusetts: Academic Press. 2011:123-148

[80] Ji Q, Zhang HY, Geng JB. What drives natural gas prices in the United States? A directed acyclic graph approach. Energy Economics. 2018;**69**:79-88

[81] Kern JD, Hise AM, Characklis GW, Gerlach R, Viamajala S, Gardner RD. Using life cycle assessment and techno-economic analysis in a real options framework to inform the design of algal biofuel production facilities. Bioresource Technology. 2017;**225**:418-428

[82] Phlips EJ, Zeman C, Hansen P. Growth, photosynthesis, nitrogen fixation and carbohydrate production by a unicellular cyanobacterium, Synechococcus sp. (Cyanophyta). Journal of Applied Phycology. 1989;**1**(2):137-145

[83] Jorquera O, Kiperstok A, Sales EA, Embirucu M, Ghirardi ML. Comparative energy life-cycle analyses of microalgal biomass production in open ponds and photobioreactors. Bioresource Technology. 2010;**101**(4):1406-1413

[84] Raes EJ, Isdepsky A, Muylaert K, Borowitzka MA, Moheimani NR. Comparison of growth of Tetraselmis in a tubular photobioreactor (Biocoil) and a raceway pond. Journal of Applied Phycology. 2014;**26**(1):247-255

[85] Narala RR, Garg S, Sharma KK, Thomas-Hall SR, Deme M, Li Y, et al. Comparison of microalgae cultivation in photobioreactor, open raceway pond, and a two-stage hybrid system. Frontiers in Energy Research. 2016;**4**:29

[86] Chisti Y. Large-scale production of algal biomass: Raceway ponds. In: Algae Biotechnology. Champions: Springer; 2016. pp. 21-40

[87] Borowitzka MA, Moheimani NR. Sustainable biofuels from algae. Mitigation and Adaptation Strategies for Global Change. 2013;**18**(1):13-25

[88] Pienkos PT, Darzins AL. The promise and challenges of microalgal-derived biofuels. Biofuels, Bioproducts and Biorefining: Innovation for a Sustainable Economy. 2009;**3**(4):431-440

[89] Jones SB, Zhu Y, Anderson DB, Hallen RT, Elliott DC, Schmidt AJ, et al. Process Design and Economics for the Conversion of Algal Biomass to Hydrocarbons: Whole Algae Hydrothermal Liquefaction and Upgrading. Richland, WA (United States): Pacific Northwest National Lab (PNNL); 2014

[90] United States Department of Agriculture (USDA). National Agricultural Statistics Service. Land Values 2017 Summary [Internet]. 2017. Available from: https://www.usda.gov/nass/PUBS/TODAYRPT/land0817.pdf [Accessed: 2018-12-05]

[91] Brown RC, Brown TR. Biorenewable Resources: Engineering New Products from Agriculture. Ames, IA: Blackwell Publishing Co.; 2003

[92] McKeown RM, Hughes D, Collins G, Mahony T, O'Flaherty V. Low-temperature anaerobic digestion for wastewater treatment. Current Opinion in Biotechnology. 2012;**23**(3):444-451

[93] Bialek K, Cysneiros D, O'Flaherty V. Low-temperature (10 C) anaerobic digestion of dilute dairy wastewater in an EGSB bioreactor: Microbial community structure, population dynamics, and kinetics of methanogenic populations. Archaea. 2013;**2013**:1-10

[94] Gunnigle E, Siggins A, Botting CH, Fuszard M, O'Flaherty V, Abram F. Low-temperature anaerobic digestion is associated with differential methanogenic protein expression. FEMS Microbiology Letters. 2015;**362**(10):fnv059

[95] United States Environmental Protection Agency (EPA). Wastewater Technology Fact Sheet EPA Anaerobic Lagoons [Internet]. 2002. Available from: https://nepis.epa.gov/Exe/ZyPDF.cgi/P100IL5S.PDF?Dockey=P100IL5S.PDF [Accessed: 2018-12-05]

[96] Hassan SN, Douglas PL, Croiset E. Techno-economic study of CO_2 capture from an existing cement plant using MEA scrubbing. International Journal of Green Energy. 2007;**4**(2):197-220

[97] Cozma P, Wukovits W, Mămăligă I, Friedl A, Gavrilescu M. Analysis and modelling of the solubility of biogas components in water for physical absorption processes. Environmental Engineering & Management Journal (EEMJ). 2013;**12**(1):1

[98] Gubicza K, Nieves IU, Sagues WJ, Barta Z, Shanmugam KT, Ingram LO. Techno-economic analysis of ethanol production from sugarcane bagasse using a liquefaction plus simultaneous saccharification and co-fermentation process. Bioresource Technology. 2016;**208**:42-48

[99] Klassen V, Blifernez-Klassen O, Hoekzema Y, Mussgnug JH, Kruse O. A novel one-stage cultivation/fermentation strategy for improved biogas production with microalgal biomass. Journal of Biotechnology. 2015;**215**:44-51

[100] Heaven S, Milledge J, Zhang Y. Comments on 'Anaerobic digestion of microalgae as a necessary step to make microalgal biodiesel sustainable'. Biotechnology Advances. 2011;**29**(1):164-167

[101] Huesemann M, Crowe B, Waller P, Chavis A, Hobbs S, Edmundson S, et al. A validated model to predict microalgae growth in outdoor pond cultures subjected to fluctuating light intensities and water temperatures. Algal Research. 2016;**13**:195-206

[102] Zhou X, Yuan S, Chen R, Song B. Modelling microalgae growth in nitrogen-limited continuous culture. Energy. 2014;**73**:575-580

[103] Blanken W, Postma PR, de Winter L, Wijffels RH, Janssen M. Predicting microalgae growth. Algal Research. 2016;**14**:28-38

[104] Ji F, Liu Y, Hao R, Li G, Zhou Y, Dong R. Biomass production and nutrients removal by a new microalgae strain Desmodesmus sp. in anaerobic digestion wastewater. Bioresource Technology. 2014;**161**:200-207

[105] Ji F, Zhou Y, Pang A, Ning L, Rodgers K, Liu Y, et al. Fed-batch cultivation of Desmodesmus sp. in anaerobic digestion wastewater for improved nutrient removal and biodiesel production. Bioresource Technology. 2015;**184**:116-122

[106] Wang M, Yang Y, Chen Z, Chen Y, Wen Y, Chen B. Removal of nutrients from undiluted anaerobically treated piggery wastewater by improved microalgae. Bioresource Technology. 2016;**222**:130-138

[107] Romero-Villegas GI, Fiamengo M, Fernández FA, Grima EM. Utilization of centrate for the outdoor production of marine microalgae at pilot-scale in flat-panel photobioreactors. Journal of Biotechnology. 2018;**284**:102-114

[108] Melis A. Solar energy conversion efficiencies in photosynthesis: Minimizing the chlorophyll antennae to maximize efficiency. Plant Science. 2009;**177**(4):272-280

[109] Klein BC, Bonomi A, Maciel FR. Integration of microalgae production with industrial biofuel facilities: A critical review. Renewable and Sustainable Energy Reviews. 2017;**82**:1376-1392

[110] Maeda Y, Yoshino T, Matsunaga T, Matsumoto M, Tanaka T. Marine microalgae for production of biofuels and chemicals. Current Opinion in Biotechnology. 2018;**50**:111-120

[111] Shuba ES, Kifle D. Microalgae to biofuels: 'Promising'alternative and renewable energy, review. Renewable and Sustainable Energy Reviews. 2018;**81**:743-755

[112] Marques AD, Pinto FP, Araújo OQ, Cammarota MC. Assessment of methods to pretreat microalgal biomass for enhanced biogas production. Journal of Sustainable Development of Energy, Water and Environment Systems. 2018;**6**(2):394-404

[113] Ometto F, Quiroga G, Pšenička P, Whitton R, Jefferson B, Villa R. Impacts of microalgae pre-treatments for improved anaerobic digestion: Thermal treatment, thermal hydrolysis, ultrasound and enzymatic hydrolysis. Water Research. 2014;**65**:350-361

[114] Passos F, Uggetti E, Carrère H, Ferrer I. Pretreatment of microalgae to improve biogas production: A review. Bioresource Technology. 2014;**172**:403-412

[115] Rodriguez C, Alaswad A, Mooney J, Prescott T, Olabi AG. Pre-treatment techniques used for anaerobic digestion of algae. Fuel Processing Technology. 2015;**138**:765-779

[116] Bohutskyi P, Betenbaugh MJ, Bouwer EJ. The effects of alternative pretreatment strategies on anaerobic digestion and methane production from different algal strains. Bioresource Technology. 2014;**155**:366-372

[117] Giraldo-Calderón ND, Romo-Buchelly RJ, Arbeláez-Pérez AA, Echeverri-Hincapié D, Atehortúa-Garcés L. Microalgae biorefineries: Applications and emerging technologies. DYNA. 2018;**85**(205):219-233

[118] Jagadevan S, Banerjee A, Banerjee C, Guria C, Tiwari R, Baweja M, et al. Recent developments in synthetic biology and metabolic engineering in microalgae towards biofuel production. Biotechnology for Biofuels. 2018;**11**(1):185

[119] Solé-Bundó M, Salvadó H, Passos F, Garfí M, Ferrer I. Strategies to optimize microalgae conversion to biogas: Co-digestion, pretreatment and hydraulic retention time. Molecules. 2018;**23**(9):2096

[120] Sathyaprakasan P, Kannan G. Economics of bio-hydrogen production. International Journal of Environmental Science and Development. 2015;**6**(5):352

[121] Mottet A, Habouzit F, Steyer JP. Anaerobic digestion of marine microalgae in different salinity levels. Bioresource Technology. 2014;**158**:300-306

[122] Doğan-Subaşı E, Demirer GN. Anaerobic digestion of microalgal (*Chlorella vulgaris*) biomass as a source of biogas and biofertilizer. Environmental Progress & Sustainable Energy. 2016;**35**(4):936-941

[123] Lü F, Ji J, Shao L, He P. Bacterial bioaugmentation for improving methane and hydrogen production from microalgae. Biotechnology for Biofuels. 2013;**6**(1):92

[124] Yang S, Liu Y, Wu N, Zhang Y, Svoronos S, Pullammanappallil P. Low-cost, Arduino-based, portable device for measurement of methane composition in biogas. Renewable Energy. 2019;**138**:224-229

[125] Perazzoli S, Steinmetz RL, Mezzari MP, Nunes EO, da Silva ML.

Biogas production from microalga biomass. In: Embrapa Suínos e Aves-Artigo em anais de congresso (ALICE). Simpósio Internacional Sobre Gerenciamento de Resíduos Agropecuários e Agroindustriais. Vol. 3. São Pedro, SP. Anais... São Pedro, SP: SBERA; 2013

[126] Passos F, Solé M, García J, Ferrer I. Biogas production from microalgae grown in wastewater: Effect of microwave pretreatment. Applied Energy. 2013;**108**:168-175

[127] Bohutskyi P, Bouwer E. Biogas production from algae and cyanobacteria through anaerobic digestion: A review, analysis, and research needs. In: Advanced Biofuels and Bioproducts. New York, NY: Springer; 2013. pp. 873-975

[128] Xin C, Addy MM, Zhao J, Cheng Y, Cheng S, Mu D, et al. Comprehensive techno-economic analysis of wastewater-based algal biofuel production: A case study. Bioresource Technology. 2016;**211**:584-593

[129] Milledge JJ, Heaven S. Energy balance of biogas production from microalgae: Development of an energy and mass balance model. Current Biotechnology. 2015;**4**(4):554-567

[130] Klassen V, Blifernez-Klassen O, Wibberg D, Winkler A, Kalinowski J, Posten C, et al. Highly efficient methane generation from untreated microalgae biomass. Biotechnology for Biofuels. 2017;**10**(1):186

[131] Juneja A, Murthy GS. Evaluating the potential of renewable diesel production from algae cultured on wastewater: Techno-economic analysis and life cycle assessment. AIMS Energy. 2017;**5**(2):239-257

[132] Xiang X, Ozkan A, Kelly C, Radniecki T. Importance of microalgae speciation on biogas production and nutrient recovery from anaerobic digestion of lipid-extracted microalgae biomass. Environmental Engineering Science. 2018;**35**(4):382-389

[133] Mussgnug JH, Klassen V, Schlüter A, Kruse O. Microalgae as substrates for fermentative biogas production in a combined biorefinery concept. Journal of Biotechnology. 2010;**150**(1):51-56

[134] Wang X, Nordlander E, Thorin E, Yan J. Microalgal biomethane production integrated with an existing biogas plant: A case study in Sweden. Applied energy. 2013;**112**:478-484

[135] Darvehei P, Bahri PA, Moheimani NR. Model development for the growth of microalgae: A review. Renewable and Sustainable Energy Reviews. 2018;**97**:233-258

[136] Kumar K, Ghosh S, Angelidaki I, Holdt SL, Karakashev DB, Morales MA, et al. Recent developments on biofuels production from microalgae and macroalgae. Renewable and Sustainable Energy Reviews. 2016;**65**:235-249

[137] Santos-Ballardo DU, Rossi S, Reyes-Moreno C, Valdez-Ortiz A. Microalgae potential as a biogas source: Current status, restraints and future trends. Reviews in Environmental Science and Bio/ Technology. 2016;**15**(2):243-264

[138] Wang X, Bao K, Cao W, Zhao Y, Hu CW. Screening of microalgae for integral biogas slurry nutrient removal and biogas upgrading by different microalgae cultivation technology. Scientific Reports. 2017;7(1):5426

[139] Lu D, Zhang XJ, Liu X, Zhang L, Hines M. Sustainable microalgae cultivation by using anaerobic centrate and biogas from anaerobic digestion. Algal Research. 2018;**35**:115-124

[140] Harun R, Davidson M, Doyle M, Gopiraj R, Danquah M, Forde G.

Technoeconomic analysis of an integrated microalgae photobioreactor, biodiesel and biogas production facility. Biomass and Bioenergy. 2011;**35**(1):741-747

[141] Barlow J, Sims RC, Quinn JC. Techno-economic and life-cycle assessment of an attached growth algal biorefinery. Bioresource Technology. 2016;**220**:360-368

[142] Batan LY, Graff GD, Bradley TH. Techno-economic and Monte Carlo probabilistic analysis of microalgae biofuel production system. Bioresource Technology. 2016;**219**:45-52

[143] Saharan BS, Sharma D, Sahu R, Sahin O, Warren A. Towards algal biofuel production: A concept of green bio energy development. Innovative Romanian Food Biotechnology. 2013;**12**:1

[144] Stiles WA, Styles D, Chapman SP, Esteves S, Bywater A, Melville L, et al. Using microalgae in the circular economy to valorise anaerobic digestate: Challenges and opportunities. Bioresource Technology. 2018;**267**:732-742

[145] Kavitha S, Subbulakshmi P, Banu JR, Gobi M, Yeom IT. Enhancement of biogas production from microalgal biomass through cellulolytic bacterial pretreatment. Bioresource Technology. 2017;**233**:34-43

[146] Kavitha S, Banu JR, Priya AA, Uan DK, Yeom IT. Liquefaction of food waste and its impacts on anaerobic biodegradability, energy ratio and economic feasibility. Applied Energy. 2017;**208**:228-238

[147] Kavitha S, Kannah RY, Banu JR, Kaliappan S, Johnson M. Biological disintegration of microalgae for biomethane recovery-prediction of biodegradability and computation of energy balance. Bioresource Technology. 2017;**244**:1367-1375

[148] Doan NT. Assessing and Enhancing Methane Productivity from Anaerobic Digestion Using Cyanothece BG0011 as Feedstock [Doctoral Dissertation]: University of Florida; 2017

[149] Zhang Y. Cultivation, growth optimization and modeling of a saline Cyanothece species BG0011 for production of biofuels and bioproducts [doctoral dissertation]. University of Florida; 2018

[150] Brown RC, Brown TR. Biorenewable Resources: Engineering New Products from Agriculture. Hoboken, New Jersey: John Wiley & Sons; 2013

[151] Doran PM. Bioprocess Engineering Principles. Cambridge, Massachusetts: Academic Press; 1995

[152] McCollum DL, Ogden JM. Techno-economic models for carbon dioxide compression, transport, and storage & correlations for estimating carbon dioxide density and viscosity. [Internet]. 2006. Available from: https://escholarship.org/uc/item/1zg00532 [Accessed: 2018-12-05]

Section 4

Biogas

Chapter 7

Biogas for Clean Energy

Demsew Mitiku Teferra and Wondwosen Wubu

Abstract

This chapter demonstrates a biogas renewable energy resource potential study for electric power generation from easily available biogas feedstock materials in four selected case study sites. Under this study, the site used in the model is a rural Kebele in Jama Woreda at 10.548° N, 39.33° E. The common biogas feedstocks considered under this study are animal slurry, human feces and jatropha byproducts whereas the biodiesel is considered from jatropha seed.

Keywords: anaerobic digestion, bioenergy, *biogas digester*, feedstock, Jatropha

1. Introduction

Biogas is a byproduct of biomass which contains methane (CH_4) and carbon dioxide (CO_2) as a main gas component in a 3:2 ratio and it is produced through micro bacterial digestion processes under anaerobic conditions from a variety of organic material from animal, agricultural, industrial and domestic wastes [1]. The biogas production level is depending on the ingredient level in the feedstock. For example; if the material consists of mainly carbohydrates, like glucose and other simple sugars and high-molecular polymers such as cellulose and hemicelluloses, the methane production is low. However, if the fat content is high, the methane production is likewise high (**Table 1**) [2].

Methane and other additional hydrogen compounds make up the combustible part of biogas. Methane is a colorless and odorless gas with a boiling point of −162°C and it burns with a blue flame. At normal temperature and pressure, methane has a density of approximately 0.75 kg/m^3. Due to carbon dioxide being somewhat heavier, biogas has a slightly higher density of 1.15–1.25 kg/m^3. Pure methane has an upper calorific value of 39.8 MJ/m^3 (11.06 kWh/m^3) (**Table 2**) [2].

Gas	%
Methane (CH_4)	55 – 70
Carbon dioxide (CO_2)	30 – 45
Hydrogen sulphide (H_2S)	1 – 2
Hydrogen (H_2)	
Ammonia (NH_3)	
Carbon monoxide (CO)	trace
Nitrogen (N_2)	trace
Oxygen (O_2)	trace

Table 1.
Biogas composition.

Substrate	HRT (days)	Solid concentration (%)	Temperature (°C)	Biogas yield (m^3/kg VS)	Methane (%)
Sewage sludge	25	6	35	0.52	68
Domestic garbage	30	5	35	0.47	—
Piggery waste	20	6.5	35	0.43	69
Poultry waste	15	6	35	0.5	69
Cattle waste	30	10	35	0.3	58
Canteen waste	20	10	30	0.6	50
Food-market waste	20	4	35	0.75	62
Mango processing waste	20	10	35	0.45	52
Tomato-processing waste	24	4.5	35	0.63	65
Lemon waste	30	4	37	0.72	53
Citrus waste	32	4	37	0.63	62
Banana peel	25	10	37	0.60	55
Pineapple waste	30	4	37	0.37	60
Mixed feed of fruit waste	20	4	37	0.62	50

Table 2.
Potential biogas production from various biomass feedstocks on VS based.

2. The biogas production process

Anaerobic digestion (AD) is a biochemical process during which complex organic matter is decomposed in absence of oxygen, by various types of anaerobic microorganisms. The result of the AD process is the biogas and the digestate. Biogas is a combustible gas, consisting primarily of methane and carbon dioxide. Digestate is the decomposed substrate, resulted from the production of biogas. If the substrate for AD is a homogenous mixture of two or more feedstock types (e.g., animal slurries and organic wastes from food industries), the process is called "co-digestion" and is common to most biogas applications today.

The process of biogas formation is a result of linked process steps, in which the initial material is continuously broken down into smaller units. Specific groups of micro-organisms are involved in each individual step. The simplified diagram of the AD process, shown in **Figure 1**, highlights the four main process steps: hydrolysis, acidogenesis, acetogenesis, and methanogenesis. The process steps quoted in **Figure 1** run parallel in time and space, in the digester tank. During hydrolysis, relatively small amounts of biogas are produced. Biogas production reaches its peak during methanogenesis [3].

Methanogenesis is a critical step in the entire anaerobic digestion process, as it is the slowest biochemical reaction of the process. Methanogenesis is severely influenced by operation conditions. Composition of feedstock, feeding rate,

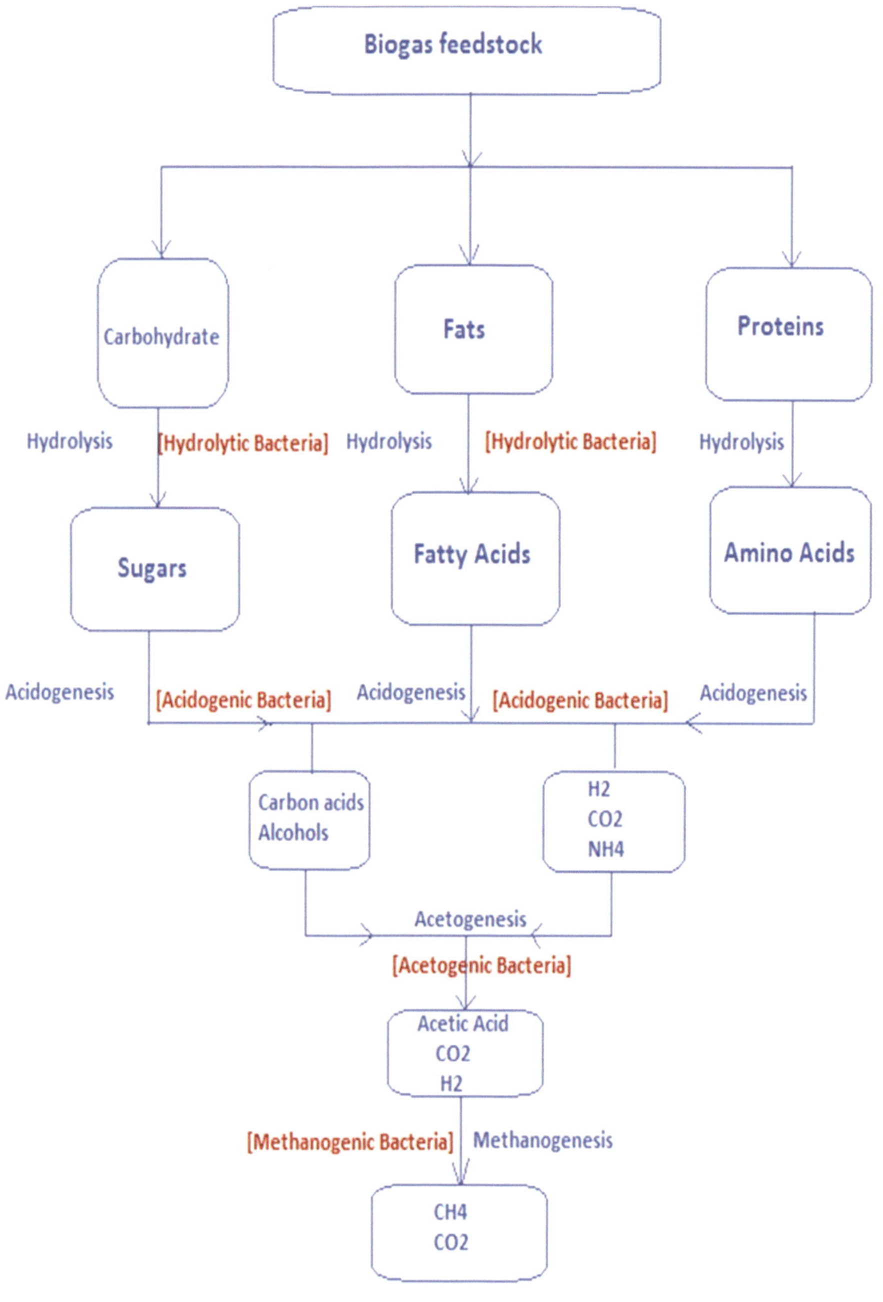

Figure 1.
Biogas production process by anaerobic digestion.

temperature, water content, NH_3 concentration and pH are examples of factors influencing the methanogenesis process.

Temperature for fermentation will greatly affect biogas production. The AD process can take place at different temperatures, divided into three temperature ranges: psychrophilic (below 20°C), mesophilic (30–42°C), and thermophilic (43–55°C). There is a direct relation between the process temperature and the HRT. The biogas production rate increases with increase the process temperature (**Table 3**).

Thermal stage	Process Temperature	Minimum HRT
Psychrophilic	< 20° C	70–80 days
Mesophilic	30–42° C	30–40 days
Thermophilic	43–55° C	15–20 days

Table 3.
Biogas production thermal stage and their corresponding retention time [4].

In practice most modern biogas plants operate at thermophilic process temperatures because this process provides many advantages, compared to mesophilic and psychrophilic processes:

- Effective destruction of pathogens
- Fast grow rate of methanogenic bacteria at higher temperature
- Minimization of biogas production period, making the process faster and more efficient
- Improve digestibility and availability of substrates
- better decomposition and utilization of solid substrates
- Increase the chance to separate liquid and solid fractions

The metabolic processes in the production of biogas from different biomass feedstocks are hydrolysis, acidogenesis, acetogenesis and methanogenesis and their byproducts in the process is represented in the figure below.

In this study thermophilic biogas temperature process is chosen in order to get higher biogas output and to achieve this target flat plate collector can be used to maintain digester process temperature at 55°c.

3. Biogas plant

A biogas plant is a complex installation, consisting of a variety of elements. The layout of such a plant depends to a large extent on the types and amounts of feedstock supplied. Now there are several main types of biogas plants all over the world. Each time it is necessary to find the most suitable type in different case. Public acceptance, cost and energy efficiency are the main criteria to install biogas plant and efficiently utilize the biogas production. In smaller areas with scarcity of biogas feedstock or slurry to use low cost clay, concert or stone masonry made biogas digester.

Installation and operation of a biogas plant is a combination of environmental, safety, economic and technical considerations. Acquiring maximum methane output, by complete digestion of feedstock substrate, would require a long fermentation or digestion time of the material inside the biogas digester and a correspondingly large digester size. The ultimate goal of biogas production is getting the highest possible methane output and having justifiable plant economy. Biogas plants have the following main components and operate with four different process stages [3].

Process stages of biogas production:

- Transport, delivery, storage and pre-treatment of feedstocks
- Biogas production
- Storage of digestate, conditioning and utilization
- Storage of biogas, conditioning and utilization.

Main components of biogas plant:

- Feedstock pre-storage tank
- Substrate mixing Tank
- Biogas digester
- Post storage tank
- Gas holder tank and
- CHP system

The amount and type of available feedstock can determine the size, type and design structure of the biogas plant. The amount of biogas feedstock could determine the dimensioning of the digester size, storage capacities and CHP unit (**Figure 2**).

The CHP system utilizes the biogas either in heat or electrical energy. The properties of the combustible methane gas (like as shown in **Table 4**) will affect the operation of the CHP equipment. The combustion nature of the gas must be

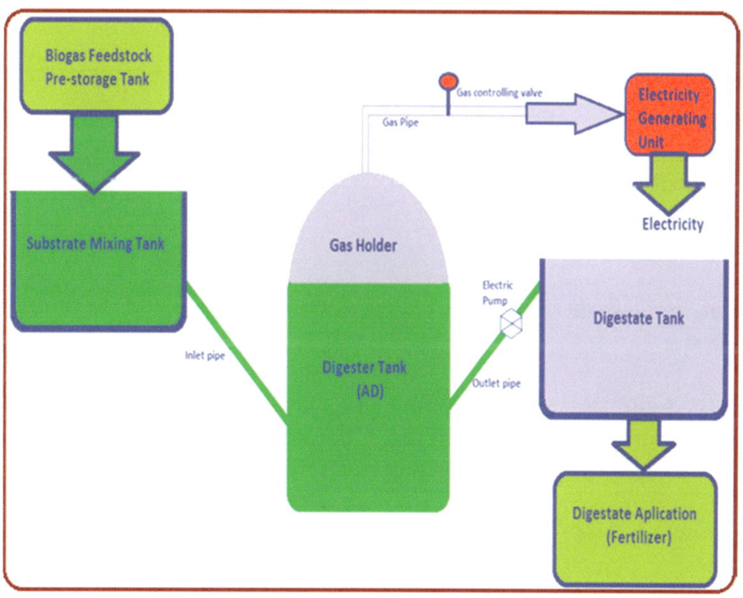

Figure 2.
Main components and general process flow of biogas production.

No.	Parameter	Symbol	Value
1.	Lower heat value	LHV	≥4 kWh/m^3
2.	Sulfur content	S	≤2.2 g/m^3 CH_4
3.	Hydrogen sulfide	H_2S	≤0.15 Vol. %
4.	Chlorine content	Cl	≤100 mg/m^3 CH_4
5.	Fluoride content	F	≤50 mg/m^3 CH_4
6.	Dust (3–10 μm)	—	≤10 mg/m^3 CH_4
7.	Relative humidity	ϕ	<90%
8.	Flow pressure	P_{gas}	20–100 mbar
9.	Gas pressure fluctuation	—	<±10% of set value
10.	Gas temperature	T	10–50°c
11.	Hydro carbon	HC	<0.4 mg/m^3 CH_4
12.	Silicon	Si	<10 mg/CH_4

Table 4.
Biogas minimum requirement used in an electric engine [3].

guaranteed, to prevent damage to the engines. Further treatment and enhancing chemical and physical properties of biogas even possible to use it for other utilizations like as vehicle fuel or in fuel cells application.

4. Design of the biogas plant

The design of the biogas plant includes the design of:

- The digester
- The gas Holder
- Digester heat maintaining system
- Siting of biogas plant

To calculate the scale of a biogas plant, certain characteristic parameters are used. These are:

- **Daily fermentation slurry feeding (Sd)**, which is an equal mixture of biogas feedstock (animal dung, human feces, poultry waste and jatropha byproduct) with water feed in to the biogas digester.
- **Retention time (RT),** the time by which the fermentation slurry stays in the digester. It is about 2–5 weeks.
- **Digester loading (R).** This parameter indicates the amount of biogas feedstock material per day is fed to the digester or to be digested. It can be measured in kg/m^3/day.

- **Specific gas production per day (Gd),** which depends on the retention time, the digestion temperature and the feed material.

4.1 Sizing of biogas digester and gasholder

The size of the digester—the digester volume (V_D)—is determined by the length of the retention time (RT) and by the amount of fermentation slurry supplied daily (S_D). The amount of fermentation slurry consists of the feed material considered in this study (e.g., cattle dung) and the mixing water.

4.1.1 Sizing of site-A biogas digester and gasholder

Daily average collectable biogas feedstock potential from cow dung, oxen dung, donkey, mule, and horse waste, chicken waste, human feces and jatropha byproduct in this study in tons/day is 10.867 = 10,867 kg/day = 15.53 m^3/day. Since the average density of animal slurry mix is 700 kg/m^3.

Additional 15.53 m^3/day water is required for proper digestion of biogas feedstock material to enhance biogas production.

HRT = 20 day, under thermophilic digestion temperature (55°C) the hydraulic retention time of the digestion process becomes short.

The volume of digester should be, $V_D = HRT \times S_D$.

= 20 day × (15.53 × 2 m^3/day) = 621 m^3.

Therefore the size of the digester for site A could be 621 m^3.

Where, V_D = the size of the digester, HRT = hydraulic retention time, and S_D is the amount of fermentation slurry (water + feedstock) feed in to the digester per day. Biogas yield in m^3/kg of fresh biogas feedstock mix is 1736.4 m^3/31850 kg = 0.054 m^3/kg; the biogas production rate is 10,867 kg/day × 0.054 m^3/kg = 588 m^3/day. Therefore the size of gasholder should account this daily biogas production.

4.1.2 Sizing of site-B biogas digester and gasholder

Daily average collectable biogas feedstock potential from cow dung, oxen dung, donkey, mule, and horse waste, chicken waste, human feces and jatropha byproduct of Site-B in tons/day is 9.253 = 9253 kg/day = 13.22 m^3/day. Since the average density of animal slurry mix is 700 kg/m^3.

Additional 13.22 m^3/day water is required for proper digestion process of biogas feedstock material to enhance biogas production.

HRT = 20 day, under thermophilic digestion temperature the hydraulic retention time of the digestion process becomes short.

The volume of digester should be, $V_D = HRT \times S_D$.

= 20 day × (13.22 × 2 m^3/day) = 529 m^3. Therefore the size of the digester for **site-B** is 529 m^3. The biogas gas production rate is 9253 kg/day × 0.054 m^3/kg = 501 m^3/day. Therefore the size of gasholder should account this daily biogas production.

4.1.3 Sizing of site-C biogas digester and gasholder

Daily average collectable biogas feedstock potential from cattle dung, donkey, mule, and horse waste, chicken waste, human feces and jatropha byproduct of site-C in tons/day is 8.82 = 8820 kg/day = 12.6 m^3/day, Since the average density of animal slurry mix is 700 kg/m^3.

Additional 12.6 m^3/day water is required for proper digestion of biogas feedstock material to enhance biogas production.

The volume of digester should be, V_D = HRT × S_D, HRT = 20 day.

= 20 day × (12.6 × 2 m^3/day) = 504 m^3.

Therefore the size of the digester for **site-C** is 504 m^3.

The gas production rate is 8820 kg/day × 0.054 m^3/kg = 477 m^3/day. Therefore the size of gasholder should account this daily biogas production also.

4.1.4 Sizing of site-D biogas digester and gasholder

Daily average collectable biogas feedstock potential of Site-D in tons/day is 3.091 = 3091 kg/day = 4.42 m^3/day, since the average density of animal slurry mix is taken as 700 kg/m^3. Additional 4.42 m^3/day water is required.

The volume of digester should be, V_D = HRT × S_D, HRT = 20 day.

= 20 day × (4.42 × 2 m^3/day) = 179 m^3.

Therefore the size of the digester for **site-D** is 179 m^3.

The gas production rate is 3091 kg/day × 0.054 m^3/kg = 168 m^3/day. Therefore the size of gasholder should account this daily biogas production.

4.2 Location of biogas plant

The next planning step in a biogas plant project idea is to find a suitable site for the establishment of the plant. The list below shows some important considerations to be made, before choosing the location of the plant: [3].

- The site should be located at suitable distance from residential areas in order to avoid inconveniences, nuisance and thereby conflicts related to odors and increased traffic to and from the biogas plant.
- The direction of the dominating winds must be considered in order to avoid wind born odors reaching residential areas.
- The site should have easy access to infrastructure such as to the electricity grid, in order to facilitate the sale of electricity and to the transport roads in order to facilitate transport of feedstock and digestate.
- The soil of the site should be investigated before starting the construction.
- The chosen site should not be located in a potential flood affected area.
- The size of the site must be suitable for the activities performed and for the amount of biomass supplied.
- The site should be located relatively close (central) to the agricultural feedstock production (manure, slurry, energy crops) aiming to minimize distances, time and costs of feedstock transportation.
- For cost efficiency reasons, the biogas plant should be located as close as possible to potential users of the produced heat and electricity.

The required site space for a biogas plant cannot be estimated in a simple way. Experience shows that for example a biogas plant of 500 kW_{el} needs an area of approximate 8000 m^2. This figure can be used as a guiding value only, as the actual area also depends on the chosen technology [3]. Based on the above criteria of site selection of biogas plant, the location of the biogas plant for each site of the study area is chosen and the detail of it is found in the economic analysis section of the biogas plant in this paper.

5. Biogas potential

5.1 Biogas potential from jatropha

Various literatures show that methane yield of jatropha fruit hull is 0.438 m^3/kg VS, and the VS is 76% of the TS of the jatropha fruit hull. Methane is 50% of the total biogas yield (1.153 m^3/kg). The biogas yield of Jatropha seed presscake is approximately 1 m^3/kg of presscake. The biogas yield of jatropha fruit hull is better than the seedcake [5]. Based on the jatropha fact sheet given in **Table 5**, the

Parameter	Unit	Minimum	Average	Maximum	Source
Seed yield	dry ton/ hectare/year	0.3	3.15	6	Position Paper on Jatropha Large Scale Project Development, FACT 2007
Fruit hull yield	dry ton/ hectare/year	0.2	2.1	4	
Rainfall requirements for seed production	mm/year	600	1000	1500	Position Paper on Jatropha Large Scale Project Development, FACT 2007
Oil content of seeds	% of mass	_	34%	40%	Jatropha bio-diesel production and use, W. Achten et al., 2008
Oil yield after pressing	% of mass of seed input	20%	25%	30%	Jatropha handbook, 2010
Presscake yield after pressing	% of mass of seed input	70	75	80	
Energy content of Seed	MJ/kg	—	37	—	

Table 5.
Jatropha fact sheet.

Biogas feedstock	Jatropha biomass, tons/year	Average jatropha biomass, tons/year	Biogas yield, m^3/kg	Methane yield, m^3/kg	Total biogas yield, m^3	Average biogas yield, m^3/year	Average methane yield, m^3/year
Presscake	4.2–96	50.1	1	0.5–0.6	4200–96,000	50,100	25,050-30,060
Fruit hull	4–80	42	1.153	0.576–0.69	4612–92,240	48,426	27,894–33,414
Total	8.2–176	92.1	1.07	0.575–0.689	8812–188,240	98,526	52,944–63,474

*Jatropha biomass (from presscake) = seed yield (ton/hectare) × % of presscake yield during oil production * total land for Jatropha farming (hectare)*
Jatropha biomass (from fruit hull) = hull yield (ton/hectare) × total land for Jatropha farming (hectare).

Table 6.
Jatropha byproduct biomass potential in the study area.

Profile	Jatropha biomass, tons	Biogas yield, m^3/kg	Biogas yield, m^3	Methane yield, m^3/kg	Methane yield, m^3
Yearly average	92.1	1.07	98,526	0.575–0.689	52,944–63,474
Daily average	0.253	1.07	270	0.575–0.689	145–174

Table 7.
Jatropha biogas potential of the study area.

Jatropha product	Jatropha oil (liter/year)	Jatropha biogas (m^3/year)	Jatropha fertilizer (kg/year)	Jatropha biomass (ton/year)
Product yield	16,090–18,774	98,526	18,420	92.1

Table 8.
Summary of Jatropha potential of the study area.

biomass, biogas and methane yield potential of the jatropha byproduct is estimated in **Tables 6, 7** and **8**.

5.2 Biogas energy potential of the study area from animal dung

A wide range of biomass types can be used as substrates (feedstock) for the production of biogas from AD. The most common biomass categories used in biogas production are listed in **Table 9** for this thesis work. To produce biogas from animal manure first we have to check whether we have animal livestock potential sufficient for biogas feedstock production or not. The following Table demonstrates the animal livestock potential for each sites of the study area.

The average fresh manure obtained from, cattle is 4.5 kg/day/head [1, 6, 7], donkey, horse and mule is 10 kg/day/head [6, 7], sheep and goat 1 kg/day/head [6, 7], and chicken is 0.08 kg/day/head [6, 7]. The average biogas yield of cattle, horse, mule, and donkey manure is 0.24 m^3/kg DM [2, 3, 8] and pigs, sheep and goat is 0.37 m^3/kg DM whereas chicken is 0.4 m^3/kg of DM [2, 3, 8]. The dry matter

Animal livestock	Site-A	Site-B	Site-C	Site-D	Ave. no. of animal/HH	Total livestock in the study area
Cows	666	566	535	172	1.7	1935
Oxen	719	612	577	184	1.85	2092
Goats	163	139	131	43	0.42	476
Sheep	1841	1567	1477	472	4.72	5350
Mule	12	10	9	3	0.03	29
Chickens	2340	1992	1878	600	6	6810
Pigs	0	0	0	0	0	0
Horse	48	40	37	12	0.12	133
Donkey	345	295	278	89	0.89	1007

Source: Jama Woreda rural development and Kebele-8 administration office, Nov 2012.

Table 9.
Jama Woreda, Kebele-8 districts animal livestock potential.

Biomass source	Average fresh manure, kg/day/head	m^3 biogas/kg DM	DM % fresh manure	Methane % biogas
Cattle	4.5	0.24	16.7	65
Pigs	2	0.37	4.4	65
Sheep, goats	1	0.37	30.7	65
Chickens	0.08	0.40	30.7	65
Horse, mule	10	0.24	7	65
Donkey	10	0.24	15	65

Total fresh manure potential of the study area (tons/day) = Average fresh manure (kg/day/head) × Total no. of livestock in study area.
Total dry mater (DM) from fresh manure = DM % of fresh manure × Total fresh manure potential of the study area (tons/day).
Total biogas production, m^3/day = Biogas m^3/kg of DM × Total dry mater (DM) from fresh manure in kg/day.
Total electricity production in kWh/day = electricity production by biogas generator from 1 m^3 biogas in kWh × total biogas production in m^3/day.
By using biogas generator it is possible to generate 1kWh electricity from 0.7 m^3 biogas [42].

Table 10.
Summary of fresh manure, biogas and methane yield of animal livestock.

Animal livestock	Ave. fresh manure, kg/day/head	Total no. of livestock in study area	Total fresh manure (ton/day)	Total DM (kg/day)	Biogas, m^3/kg of DM	Total biogas, m^3/day	Electricity production, kWh/day
Cows	4.5	1935	8.708	1455	0.24	350	500
Oxen	4.5	2092	9.414	1573	0.24	378	540
Goats	1	476	0.476	147	0.37	55	79
Sheep	1	5350	5.350	1643	0.37	608	869
Mule	10	29	0.290	24	0.24	6	9
Chicken	0.08	6810	0.545	168	0.40	68	98
Pigs	2	0	0.000	0.00	0.37	0.0	0.0
Horse	10	133	1.330	92	0.24	22	32
Donkey	10	1007	10.070	1511	0.24	363	519
Total animal manure biomass			**36.183**	**6613**	**0.28**	**1850**	**2646**

Table 11.
Summary of expected animal manure potential of the study area.

content from the total mass of fresh animal manure and the proportion of methane from the total biogas production is summarized in **Table 10** [2, 3, 9] (**Table 11**).

For a given size of plant (rated gas production capacity per day) the amount of feedstock required can be estimated using the biogas yield data provided. The specific biogas consumption in biogas engines is 0.6–0.8 m^3/kWh [1]. This specific fuel consumption value can be used to calculate the requirement for biogas for power generation purposes. The expected biomass potential from animal manure of the case study area is 36.2 tons/day and its biogas production capacity is 1850 m^3/day. Various literatures show that the collection efficiency of animal manure varies from country to country and region to region.

Most significantly the collection efficiency varies from 50 to 100% [10]. Let as consider collection efficiency of 90% for cattle, donkey, mule, horse, pig and chicken manure, 50% for goat and sheep manure and 100% for human feces based

Animal livestock	Ave. fresh manure, kg/day/ head	Total no. of livestock in study area	Total collectable fresh manure, tons/day	Total collectable DM, kg/day	Biogas, m^3/kg of DM	Total biogas, m^3/day	Electricity production, kWh/day
Cows	4.5	1935	7.837	1309.5	0.24	315	450
Oxen	4.5	2092	8.473	1415.7	0.24	340	486
Goats	1	476	0.238	73.5	0.37	27.3	39
Sheep	1	5350	2.675	821.5	0.37	304	434.3
Mule	10	29	0.261	21.6	0.24	5.2	7.43
Chicken	0.08	6810	0.491	151.2	0.40	60.5	86.43
Pigs	2	0	0.000	0.00	0.37	0.0	0.0
Horse	10	133	1.197	82.8	0.24	19.9	28.43
Donkey	10	1007	9.063	1360	0.24	326.4	466.3
Total animal manure Biomass			30.235	5235.8	0.27	1398.3	1998

Table 12.
Summary of collectable animal manure potential of the study area.

on their difficulty of collecting it. Therefore the biomass potential available for biogas generation is estimated as follows.

The total collectable fresh animal manure biomass potential of the study area is estimated to be 30.235 tons/day and its biogas production capacity is 1398.3 m^3/day (**Table 12**).

5.3 Biogas potential of the study area from human feces

Human feces are another feedstock for biogas production in the study area and the potential biogas production from human feces is discussed in this section. Feces are mostly made of water (about 75%). The rest is made of dead bacteria that helped us digest our food, living bacteria, protein, undigested food residue (known as fiber), waste material from food, cellular linings, fats, salts, and substances released from the intestines (such as mucus) and the liver (**Table 13**).

One person produces on average 100–140 g of feces per day, the dry matter content of which is about 25% and its biogas yield of about 0.2 m^3/kg DM [11]. The total collectable fresh manure biomass potential of the case study area from humans is estimated to be 0.681 tons/day and its biogas production capacity is 34.05 m^3/day. This figure accounts the collection efficiency of human excreta. **Table 14** demonstrates the biogas potential of the study area from human feces.

Population	Site-A	Site-B	Site-C	Site-D	Total
Number of household	390	332	313	100	1135
Average Family per household	4.39 (5)	4.39 (5)	4.39 (5)	4.39 (5)	4.39 (5)
Total population	1950	1660	1565	500	5675

Table 13.
Jama Woreda, Kebele-8 districts population data.

Live stock	Ave. fresh manure, kg/day/head	Total no. of population	Total fresh manure potential (ton/day)	Total DM (kg/day)	Biogas, m^3/kg DM	Total biogas, m^3/day	Electricity production, kWh/day
Human	0.12	5675	0.681	170.25	0.2	34.05	48.7

Table 14.
Biogas potential of study area from human feces.

5.4 Total biogas potential of the study area

The total biogas potential from Jatropha byproduct, Animal waste and human feces discussed above can be summarized in this section.

> **Taking** *the density of biogas 1.15 kg/m^3 and calculating the gasification ratio (the mass of biogas produced per unit mass of feed stock consumed) of the biogas system. From* ***Table 15*** *the mass of biogas feedstock consumed is 31,850 kg/day and the gas produced is 1736.4 m^3/day. Therefore the gasification ratio of biogas feedstock mix is 1736.4 m^3/31850 kg = 0.0545 m^3/kg =* **0.0626 kg/kg.**

As we have seen from **Table 15**, animal manure is the major biogas feedstock constitutes which accounts 97% from the total biogas feedstock potential whereas jatropha byproducts and human excreta constitute 1 and 2% of the total biogas feedstock potential of the study area respectively. However, the share of biogas production from, animal manure is 82%, and human excreta is 2% but biogas production from jatropha byproduct is increase to 16% regardless of its low contribution to the biomass potential since the biogas yield of jatropha byproduct is high as compared to both animal and human manure and this can be summarized in **Figure 3** given below.

Animal Livestock	Ave. fresh manure, kg/day/head	Total no. of live stock	Total collectable fresh manure (ton/day)	Total collectable DM (kg/day)	Biogas, m^3/kg DM	Total biogas production, m^3/day	Electricity yield, kWh/day
Cows	4.5	1935	7.837	1309.5	0.24	315	450
Oxen	4.5	2092	8.473	1415.7	0.24	340	486
Goats	1	476	0.238	73.5	0.37	27.3	39
Sheep	1	5350	2.675	821.5	0.37	304	434.3
Mule	10	29	0.261	21.6	0.24	5.2	7.43
Chicken	0.08	6810	0.491	151.2	0.40	60.5	86.43
Pigs	2	0	0.000	0.00	0.37	0.0	0.0
Horse	10	133	1.197	82.8	0.24	19.9	28.43
Donkey	10	1007	9.063	1360	0.24	326.4	466.3
Human	0.12	5675	0.681	170.25	0.2	34.05	48.7
Jatropha byproduct biomass			0.253	253	1.07	270	386
Total			31.85	5829.3	0.3	1736.4	2481.4

Table 15.
The total biogas and collectable feedstock potential of the study area.

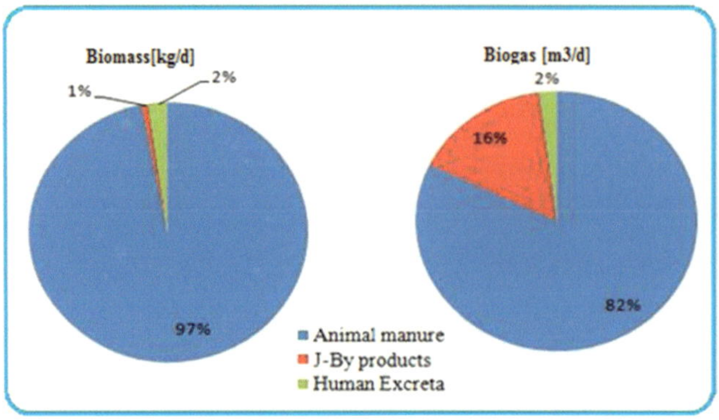

Figure 3.
Biogas feedstock contributions for biogas production in the study area.

5.5 Monthly variation of the biogas feed stock potential

The variation of jatropha byproduct feedstocks is assumed to be constant throughout the year and the potential biomass obtained from it was divided to each site regardless of the total house hold in each of the study area.

However, the biomass obtained from animal is highly depending on the availability and type of the animal feeding material. The animal feeding materials are varying in type and amount from month to month in the study area. In June and July there is enough root grass in addition to the usual animal food, let as consider this value as the annual average in ton/day (the data obtained by multiplying the biomass obtained per animal live stock in ton/day with the total number of animal live stock for each animal group in the district), as a reference frame. In January, February, and December there is excess dry agricultural farm grass for the animal food in the study area and assuming a 5% biomass resource increment is expected from the reference. March and April is a dry season and there is no enough food for the animal so considering a 5% biomass resource decrement from the reference. May, extremely drought month and August, animal grazing area are not permitted for animal food assuming a 10% animal based biomass resource drop is expected. From September to November there is excess animal food and a 10% biomass growth is assumed. Also assuming chicken manure and human feces are constant throughout the year. Taking in to account the assumption listed above the biogas feedstock potential month to month variation is presented in **Tables 16–19**.

6. Conclusion

The renewable energy potential of the site is estimated based on the primary data collected directly from the study area and secondary data obtained from various sources. The biogas feedstock mix potential of the study area is found to be 10.9 tons/day, 9.25 tons/day, 8.81 tons/day and 3.09 tons/day for Site-A, Site-B, Site-C and Site-D respectively with a gasification ratio of 0.0626 kg/kg. The study result shows that there is a sufficient biogas feedstock potential for all districts of the study area and the feasibility simulation result demonstrates there is an excess biogas after running a biogas generator in a hybrid system. The excess biogas left unused from a hybrid electric generating unit would go to biogas cooking application for the community cooking loads. Also, the biodiesel potential of the study area from Jatropha is estimated to be 18.5 m^3/year.

Month	Biomass, tons/day									
	Cow	Oxen	Mule	Horse	Donkey	Sheep	Goats	Chicken	Jatroph	Human	Total
Jan	2.82	3.069	0.114	0.4536	3.28	0.967	0.086	0.17	0.0633	0.234	11.257
Feb	2.82	3.069	0.114	0.4536	3.28	0.967	0.086	0.17	0.0633	0.234	11.257
Mar	2.552	2.775	0.1031	0.4104	2.97	0.875	0.08	0.17	0.0633	0.234	10.233
Apr	2.552	2.775	0.1031	0.4104	2.97	0.875	0.08	0.17	0.0633	0.234	10.233
May	2.417	2.63	0.0972	0.3654	2.811	0.83	0.074	0.17	0.0633	0.234	9.693
Jun	2.686	2.921	0.108	0.432	3.123	0.921	0.082	0.17	0.0633	0.234	10.740
Jul	2.686	2.921	0.108	0.432	3.123	0.921	0.082	0.17	0.0633	0.234	10.740
Aug	2.417	2.63	0.0972	0.3654	2.811	0.83	0.074	0.17	0.0633	0.234	9.6912
Sep	2.954	3.213	0.119	0.475	3.4353	1.013	0.09	0.17	0.0633	0.234	11.767
Oct	2.954	3.213	0.119	0.475	3.4353	1.013	0.09	0.17	0.0633	0.234	11.767
Nov	2.954	3.213	0.119	0.475	3.4353	1.013	0.09	0.17	0.0633	0.234	11.767
Dec	2.82	3.069	0.114	0.4536	3.28	0.967	0.086	0.17	0.0633	0.234	11.257
Average	2.693	2.958	0.1096	0.4335	3.1628	0.9327	0.083	0.17	0.0633	0.234	10.867

Table 16.
Biomass resource of site-A—390 families.

Month	Biomass, tons/day										
	Cow	Oxen	Mule	Horse	Donkey	Sheep	Goats	Chicken	Jatropha	Human	Total
Jan	2.40	2.614	0.095	0.378	2.788	0.823	0.073	0.144	0.0633	0.183	9.56
Feb	2.40	2.614	0.095	0.378	2.788	0.823	0.073	0.144	0.0633	0.183	9.56
Mar	2.17	2.364	0.086	0.342	2.523	0.744	0.066	0.144	0.0633	0.183	8.69
Apr	2.17	2.364	0.086	0.342	2.523	0.744	0.066	0.144	0.0633	0.183	8.69
May	2.056	2.240	0.081	0.324	2.390	0.706	0.062	0.144	0.0633	0.183	8.25
Jun	2.284	2.489	0.09	0.36	2.655	0.784	0.070	0.144	0.0633	0.183	9.12
Jul	2.284	2.489	0.09	0.36	2.655	0.784	0.070	0.144	0.0633	0.183	9.12
Aug	2.056	2.240	0.081	0.324	2.38	0.706	0.062	0.144	0.0633	0.183	8.24
Sep	2.513	2.737	0.099	0.469	2.921	0.862	0.077	0.144	0.0633	0.183	10.07
Oct	2.513	2.737	0.099	0.469	2.921	0.862	0.077	0.144	0.0633	0.183	10.07
Nov	2.513	2.737	0.099	0.469	2.921	0.862	0.077	0.144	0.0633	0.183	10.07
Dec	2.40	2.614	0.095	0.378	2.788	0.823	0.073	0.144	0.0633	0.183	9.56
Average	2.313	2.52	0.09	0.383	2.69	0.794	0.071	0.144	0.0633	0.183	9.25

Table 17.
Biomass resource of site-B—332 families.

DOI: http://dx.doi.org/10.5772/intechopen.79534

Month	Biomass, tons/day										
	Cow	Oxen	Mule	Horse	Donkey	Sheep	Goats	Chicken	Jatropha	Human	Total
Jan	2.263	2.46	0.085	0.755	2.637	0.776	0.069	0.136	0.0633	0.188	9.434
Feb	2.263	2.46	0.085	0.755	2.637	0.776	0.069	0.136	0.0633	0.188	9.431
Mar	2.048	2.23	0.077	0.316	2.385	0.702	0.062	0.136	0.0633	0.188	8.206
Apr	2.048	2.23	0.077	0.316	2.385	0.702	0.062	0.136	0.0633	0.188	8.206
May	1.94	2.13	0.073	0.30	2.26	0.665	0.059	0.136	0.0633	0.188	7.812
Jun	2.156	2.35	0.081	0.333	2.511	0.739	0.066	0.136	0.0633	0.188	8.618
Jul	2.156	2.35	0.081	0.333	2.511	0.739	0.066	0.136	0.0633	0.188	8.618
Aug	1.94	2.123	0.073	0.30	2.26	0.665	0.059	0.136	0.0633	0.188	7.812
Sep	2.37	2.556	0.089	0.41	2.76	0.813	0.072	0.136	0.0633	0.188	9.457
Oct	2.37	2.556	0.089	0.41	2.76	0.813	0.072	0.136	0.0633	0.188	9.457
Nov	2.37	2.556	0.089	0.418	2.76	0.813	0.072	0.136	0.0633	0.188	9.457
Dec	2.263	2.463	0.085	0.755	2.637	0.669	0.069	0.136	0.0633	0.188	9.328
Average	2.183	2.372	0.082	0.449	2.542	0.739	0.067	0.136	0.0633	0.188	8.820

Table 18.
Biomass resource of site-C—313 families.

Month	Biomass, tons/day										
	Cow	Oxen	Mule	Horse	Donkey	Sheep	Goats	Chicken	Jatropha	Human	Total
Jan	0.723	0.787	0.029	0.1134	0.841	0.496	0.044	0.0432	0.0633	0.06	3.199
Feb	0.723	0.787	0.029	0.1134	0.841	0.496	0.044	0.0432	0.0633	0.06	3.199
Mar	0.654	0.712	0.026	0.1026	0.761	0.448	0.04	0.0432	0.0633	0.06	2.910
Apr	0.654	0.712	0.026	0.1026	0.761	0.448	0.04	0.0432	0.0633	0.06	2.910
May	0.620	0.674	0.024	0.0972	0.721	0.425	0.038	0.0432	0.0633	0.06	2.766
Jun	0.689	0.750	0.027	0.108	0.801	0.472	0.043	0.0432	0.0633	0.06	3.056
Jul	0.689	0.750	0.027	0.108	0.801	0.472	0.043	0.0432	0.0633	0.06	3.056
Aug	0.620	0.675	0.024	0.0972	0.721	0.425	0.038	0.0432	0.0633	0.06	2.766
Sep	0.757	0.825	0.03	0.1188	0.881	0.519	0.046	0.0432	0.0633	0.06	3.344
Oct	0.757	0.825	0.03	0.1188	0.881	0.519	0.046	0.0432	0.0633	0.06	3.344
Nov	0.757	0.825	0.03	0.1188	0.881	0.519	0.046	0.0432	0.0633	0.06	3.344
Dec	0.723	0.787	0.028	0.1134	0.841	0.496	0.044	0.0432	0.0633	0.06	3.199
Average	0.697	0.759	0.027	0.1094	0.811	0.478	0.043	0.0432	0.0633	0.06	3.091

Table 19.
Biomass resource of site-D—100 families.

DOI: http://dx.doi.org/10.5772/intechopen.79534

Author details

Demsew Mitiku Teferra* and Wondwosen Wubu
Addis Ababa Science and Technology University, Addis Ababa, Ethiopia

*Address all correspondence to: demsewmitku@gmail.com

References

[1] Hadagu A. Status and Trends of Ethiopian Rural Electrification Fund; 2006. http://www.bgr.de/geotherm/ArGeoC1/pdf/07%20A.%20Hadgu%20status%20of%20Ethiopian%20Electrification%20fund.pdf

[2] Maithel S. Biomass Energy Resource Assessment Handbook—prepared for Asian and Pacific Centre for Transfer of Technology of the United Nations – Economic and Social Commission for Asia and the Pacific (ESCAP); September 2009. https://en.calameo.com/read/001424067e8c637721c50

[3] Jørgensen PJ. Plan Energi and researcher for a day. In: Biogas – green energy. 2nd ed. Digisource Denmark: Faculty of Agricultural Sciences, Aarhus University; 2009. http://lemvigbiogas.com

[4] Al Seadi T, Rutz D, Prassl H, Köttner M, Finsterwalder T, Volk S, Janssen R. Biogas Handbook. Niels Bohrs Vej 9-10, DK-6700 Esbjerg, Denmark: University of Southern Denmark Esbjerg; 2008

[5] www.fact-foundation.com

[6] Janske van E, André F. Global experience with jatropha cultivation for bioenergy: An assessment of socio-economic and environmental aspects. Heidelberglaan, The Netherlands: Copernicus Institute, Utrecht University, The Netherlands Eindhoven University of Technology

[7] http://home.t-online.de/home/320033440512-0002/downloads/jcl-manual.pdf

[8] Jongschaap REE et al. Claims and Facts on *Jatropha curcas* L., Global *Jatropha curcas* Evaluation, Breeding and Propagation Programme. Wageningen UR: Plant Research International; 2007

[9] Demirbas A. Conversion of biomass using glycerine to liquid fuel for blending gasoline as alternative engine fuel. Energy Conversion and Management. 2000;**41**:1741-1748

[10] Nicholson FA, Chambers BJ, Williamsb JR, Unwin RJ. Heavy Metal Contents of Livestock Feeds and Animal Manures in England and Wales. Meden Vale, Mans®eld, Nottinghamshire, NG20 9PF, UK: ADAS Gleadthorpe Research Centre; 1999

[11] Milbrandt A. Assessment of Biomass Resources in Liberia Prepared for the U.S. Agency for International Development (USAID) under the Liberia Energy Assistance Program (LEAP). Technical Report, NREL/TP-6A2-44808; April 2009. http://www.osti.gov/bridge

Chapter 8

Non-Catalytic Reforming of Biogas in Porous Media Combustion

Mario Toledo Torres

Abstract

Rich combustion of biogas inside an inert porous media reactor was investigated to evaluate hydrogen and syngas production. Temperature, velocities, and product gas composition of the combustion waves were analysed, while varying its filtration velocity, for a range of equivalence ratios (φ) from $\varphi = 1.0$ to $\varphi = 3.5$. A numerical model based on comprehensive heat transfer and chemical mechanisms was found to be in a good qualitative agreement with experimental data. Partial oxidation products of biogas (H_2 and CO) were dominant on rich combustion. Different gas mixtures of methane and carbon dioxide, which simulated synthetic biogas, and the addition of a varying fraction of water steam were experimentally analysed. It was observed that an increasing steam to carbon ratio (S/C) improved hydrogen and syngas production. The non-catalytic process investigated results in an effective biogas upgrading, and to be essentially higher than under natural gas filtration combustion.

Keywords: biogas reforming, porous media burner, hydrogen, syngas, non-catalytic

1. Introduction

Large-scale production of hydrogen (H_2) is mainly obtained through the thermochemical conversion of methane (CH_4) into H_2 and carbon monoxide (CO), a mixture also known as syngas. The main conversion processes are dry reforming, partial oxidation, steam reforming, and autothermal reforming, all of which typically use fossil fuels as main carbonaceous feedstock, the natural gas being the most widely used. However, current efforts have been focused on the development of sustainable, carbon-neutral alternatives for H_2 production. Hence, biogas upgrading by partial oxidation of CH_4 in the presence of oxygen (O_2), steam (H_2O), or carbon dioxide (CO_2) is considered an interesting alternative to produce syngas while reducing GHG emissions to the atmosphere.

Accordingly, research has been focused on improving the syngas production from biogas process efficiency, being the use of fluidized bed reactors a promising real alternative to effectively increase efficiency. However, they have presented a major drawback related to catalytic wearing mainly associated to the elevated temperatures (over 1000 K and up to 1300 K) required for the efficient and cost-effective conversion of biogas into syngas [1–4].

Moreover, the widely reported effects of sintering and coking on the catalyst both responsible for the generation of depositions that cause catalytic deactivation [5–6], observed to occur on high-temperature regimes (over 1100 K), have prevented the commercial development of the technology. Nevertheless, further studies related to upgrading the thermal properties of catalysts used in biogas reforming, and their resistance to the aforementioned phenomena, have been performed, particularly as a countermeasure to the effects of GHG [7–9].

However, since the catalytic approach has proven to be thermally challenging, an alternative non-catalytic method, such as partial oxidation (POX) in inert porous media (IPM), has proven to be an interesting option for high-temperature biogas conversion. The advantage of using a porous matrix to enhance several reaction processes, such as combustion, partial oxidation, steam reforming, and dry reforming, among others, has been extensively studied by numerous researchers, where the use of a chemically inert porous media enables the propagation of an autothermal reacting wave which benefits from the increased heat transfer on the reaction zone due to the solid matrix. Specifically, this enhanced thermal mechanism, mainly attributed to the heat conduction and radiation, and the highly developed inner surface of the porous media being heated by the reaction wave, acts as a heat recirculation mechanism which distributes the thermal energy up- and downstream of the reacting zone, thus preheating the fresh mixture and homogenising the reaction temperature across the reacting wave. Furthermore, the existence of multiple flow paths for the filtered gas increases its diffusion and heat transfer with the solid phase. Finally, filtration combustion has shown to increase the operational ranges of free-flame combustion on a wide range of filtration velocities, equivalence ratios, and power loads.

When combusting gaseous fuels in porous media, steady and transient systems are the two approaches commonly employed [10–16]. The first approach is widely used in radiant burners and surface combustor-heaters, where the combustion wave maintains its position due to an equilibrium of the heat transfer mechanisms. While the transient operation considers a reaction wave travelling on an upstream or downstream fashion through the porous media, the direction of wave propagation depends mostly on the physical properties of solid and gas, filtration velocity, temperature, and air excess of the mixture. Combining these parameters, it can be noted that the movement velocities of these waves are much lower than free-flame combustion velocities [13].

A transient operation mode is characterised by concentrating or diluting the heat released from the chemical reaction while travelling in a downstream or upstream direction, respectively; thus the reaction front can reach a temperature considerably different from the free-flame adiabatic flame temperature. This phenomenon is mainly attributed to the reaction chemistry and the heat transfer mechanisms. Under a counterflow configuration, the downstream propagation results in superadiabatic reaction waves which effectively increase the conventional free-flame flammability limits for both ultra-lean and ultrarich operation modes.

Superadiabatic filtration combustion of rich and ultrarich mixtures allows the stable operation of both partial oxidation and thermal cracking of hydrocarbons. This technology for hydrogen or syngas production uses an IPM [17–20]. The fuels used in porous combustion systems are basically of gaseous form due to fluidity, volumetric capacity, and shorter mixing length scale [21–28]. Accordingly, this chapter presents the numerical and experimental results obtained for different biogas compositions (CH_4 and CO_2) on syngas production by filtration combustion in an IPM reactor.

2. Numerical model

Experimental temperature measurements showed a minimal radial gradient; thus a one-dimensional numerical simulation was considered adequate for representing the reaction wave inside the porous media. A volume-averaged model [29] was used to solve the two-temperature mathematical model proposed to describe the filtration combustion under isobaric, stationary, and one-dimensional conditions. The combustion wave propagation rate was considered to be at least three orders of magnitude smaller than the filtration velocity of the gaseous mixture. The two-temperature approximation was formulated in order to describe a fully developed stationary reaction wave in a coordinate system moving with the reaction zone [19, 25, 30]. Therefore, this model describes both solid and gaseous phases through their fluid dynamics and heat transfer interactions.

Continuity equation.

$$\frac{\partial(\dot{M})}{\partial x} = \frac{\partial(\varepsilon\rho v)}{\partial x} = 0 \tag{1}$$

Gas phase energy equation.

$$c_p\rho_g v\frac{\partial T_g}{\partial x} = \frac{\partial}{\partial x}\left(\left[k_g + (c_p\rho_g)D_{ax}\right]\frac{\partial T_g}{\partial x}\right) - \sum_k\left(\dot{\omega}_k h_k W_k - \rho_k Y_k V_k c_{pk}\frac{\partial T_g}{\partial x}\right) - \frac{h_v}{\varepsilon}(T_g - T_s) \tag{2}$$

Solid phase energy equation.

$$-(1-\varepsilon)c_s\rho_s u\frac{\partial T_s}{\partial x} = \frac{\partial}{\partial x}\left([k_e + k_R]\frac{\partial T_g}{\partial x}\right) - \beta(T_s - T_0) - h_v(T_g - T_s) \tag{3}$$

Species conservation equation.

$$\rho_g v\frac{dY_k}{dx} + \frac{d}{dx}(\rho_g Y_k V_k) = \dot{\omega}_k W_k \tag{4}$$

Boundary conditions chosen for the inlet and outlet of the reactor consider thermal equilibrium between gas and solid phases.

Inlet:	$x = 0$,	$T_0 = T_g = T_s$,	$Y = Y_{k,0}$
Outlet:	$\frac{\partial Y_k}{\partial x} = 0$,	$T_s = T_g = T$,	$\frac{\partial T}{\partial x} = \beta(T - T_0)/(\varepsilon c_p\rho_g v - (1-\varepsilon)c_s\rho_s u)$

An analytical solution of Eqs. (1)–(4) was used to impose the boundary conditions of the reaction wave temperature at the outlet. This is achieved by assuming the aforementioned conditions for the outlet and adding that $\dot{\omega}_k = 0$ $h_v = (6\varepsilon/d^2)Nuk_g$ and that radiation can be neglected at the outlet since it is far from the reaction zone. Thus, calculation for temperature and species in the reaction wave can be computed for a finite and a well-defined spatial domain. The convective heat transfer coefficient (h_v) was taken from [19] as $h_v = (6\varepsilon/d^2)Nu\ k_g$, whereas the correlation for Nu was considered as

$Nu = 2 + 1.1Re^{0.6}Pr_g^{1/3}$ as presented by Wakao and Kaguei [31]. A radiant conductivity model taken from [29] was used to include the effect of radiation, where $k_R = 4F\sigma T_s^3$ with F being the radiation exchange factor. This has to be modelled for each material since it is dependent upon its thermal conductivity and emissivity. For the solid phase, composed of 5.6 mm in diameter alumina spheres, values from 0.3 to 0.6 are used. Effective thermal conductivity of the packed bed and its porosity were estimated as $k_e = 0.005\ k_s$ and $\varepsilon = 0.4$. The tortuosity of the porous media and its porosity contributes to flow irregularities which affect the effective mass diffusion of species in the gas phase; this phenomenon is described by an axial gas dispersion coefficient, $D_{ax} = 0.5dv$ [31]. Dispersion coefficients for both thermal and mass diffusivities are considered to be equal according to a heat/mass transfer analogy. As presented by Henneke and Ellzey [30], a sum of molecular diffusion and dispersion is used to represent the effective diffusion.

Finally, all thermophysical properties from the solid phases, such as thermal conductivity, heat capacity, and radiative properties, were obtained from openly available technical reports [32] and verified against the technical specifications from the ceramic manufacturer (Coors, Inc.).

Chemical kinetics of the process was modelled through the implementation of the GRI 3.0 [33] chemical kinetics mechanism, which includes NO_X chemistry, alongside the CHEMKIN [34] package. Even though this mechanism is not designed specifically to simulate reactions of ultra-lean nor ultrarich mixtures, it is considered as an acceptable first approach to understand the possible reaction mechanisms occurring under these conditions.

The calculations were performed for a given value of the filtration (interstitial) velocity, and the implemented numerical algorithm in the modified PREMIX [35] code was used to find the wave propagation velocity.

3. Syngas production by filtration combustion

In this section, numerical and experimental results are showed for wave velocities, combustion temperature, and H_2 and CO concentrations for rich combustion of biogas in porous media. Additionally, results on rich combustion of natural gas in porous media are presented.

The combustion system is shown in **Figure 1** and consists on a cylindrical quartz tube. This tube is filled with a bed of alumina spheres (Al_2O_3) with an average diameter of 5.6 mm, yielding a porosity of ~40%. The inner and outer surfaces of the tube were covered with a 2 and 25 mm thick Kaowool insulation material, respectively. System diagnostics were required to assess the temperature profile in the reactor along with emission concentrations in the product gases. The axial temperature distribution of the reactor was acquired using S-type thermocouples. These thermocouples were housed in a long multibore ceramic shell; therefore measured temperatures were considered to be very close to the temperatures of the solid phase. Temperature measurements were digitised using a data acquisition module and transferred to a PC. The reaction wave propagation rates were obtained from thermocouple measurements over time and the known distance of the thermocouples. The concentrations of H_2, CO, CH_4, and CO_2 in the product gases were measured using a gas chromatograph fitted with a thermal conductivity detector (TCD), while gas samples were acquired through an alumina tube immersed at the end of the reactor. To avoid the effect of external air vortices on the composition, the probe was inserted 20 mm into the packed bed.

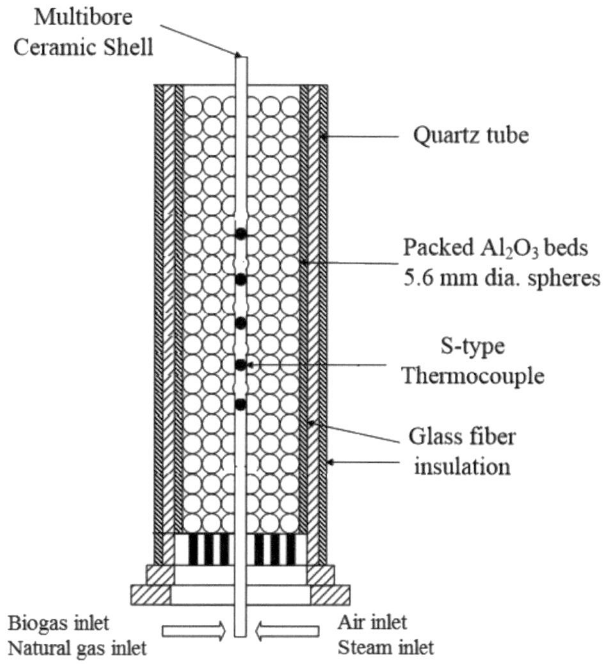

Figure 1.
Schematic of the experimental setup.

The methane, carbon dioxide, air and steam mixture flowed continuously through the quartz tube, and the reactant concentrations were controlled using Aalborg mass flow controllers. A temperature experimental measurement error was estimated at 50 K, which is considered mainly as a radial error. The reaction wave velocity measurement error was estimated at 10%, and the chemical sampling accuracy was considered close to 10%. The experimental uncertainties were defined based on the accuracy of the laboratory equipment and the repeatability of the experimental data.

3.1 Combustion wave temperature and propagation rate

One objective of the numerical simulation was to clarify the effect of interfacial heat transfer. Calculations performed with the same initial conditions in two-temperature approximations are presented in **Figure 2** for natural gas-air and biogas-air mixtures. Thermal non-equilibrium between gas and solid phases needs to be applied for consideration of propagation waves. Downstream, upstream, and standing waves were observed for natural gas-air and biogas-air mixtures, mainly depending on the equivalence ratio. Stable combustion of rich and ultrarich mixtures was observed experimentally for the region of equivalence ratios studied.

For natural air mixtures (**Figure 2A**), upstream wave propagation was observed for the range of equivalence from stoichiometric to 1.7. The velocity of the wave decreases with an increase of the natural gas concentration, approaching zero at 1.7. A stationary combustion wave is formed under these conditions. With further increase of the natural gas content, the regime of propagation changes towards a downstream direction. This regime is observed for the range of equivalence ratios from 1.7 to 3.5, where the speed increases with the natural gas content. It was found experimentally that the maximum combustion temperature remains almost constant (Ts = 1529 K) throughout the rich region and is practically independent of the natural gas content. These combustion temperatures are attributed to the changes of

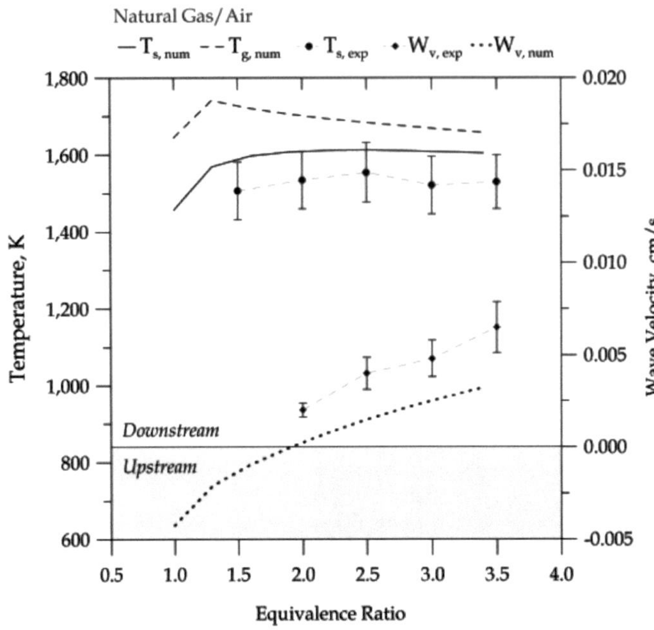

(A)

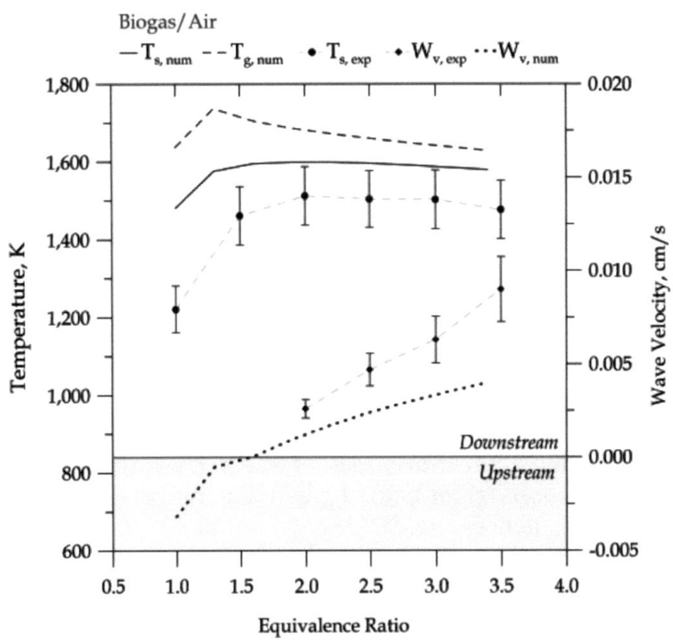

(B)

Figure 2.
Combustion temperatures and wave velocities for natural gas-air mixtures (A) and biogas (60% CH_4/40% CO_2)-air mixtures (B) varying the equivalence ratio from stoichiometry (φ = 1.0) to φ = 3.5.

combustion chemistry, because all other governing parameters such as flow rates, porous body properties, and heat content are similar.

For biogas-air mixtures, the combustion in porous media shows similar behaviour (**Figure 2B**). Upstream wave propagation is observed for the range of equivalence from stoichiometric to 1.5. A standing combustion wave was formed at 1.5. The regime of downstream propagation was observed for the range of equivalence ratios from 1.5 to 3.5. It was found experimentally that the maximum combustion temperature remains almost constant (Ts = 1491 K) throughout the rich region and was practically independent of the biogas content.

3.2 Combustion products

Starting from equivalence ratios higher than 1.0, complete combustion of natural gas and biogas will not be achieved because of the low oxygen content in the mixture. Consequently, concentrations of CO_2 and H_2O decrease, while partial oxidation products such as H_2 and CO increase their presence in the product gases (**Figure 3**).

Hydrogen and carbon monoxide concentration on the reaction waves of natural gas and biogas showed a direct relation to an increase of equivalence ratio. Unburned methane was detected in the gaseous products starting from $\varphi \approx 1.3$. Its concentration grew with an increasing equivalence ratio. Natural gas and biogas thermochemical processing could be characterised as fuel reformation or cracking rather than combustion.

In comparison, product concentration had a similar behaviour for natural gas and biogas. H_2 and CH_4 concentrations are highest for natural gas and CO_2 and CO for biogas, in the range of equivalence ratios studied, which is consistent with the previous results, under similar conditions, as reported by [10, 16, 20].

3.3 Biogas composition

Biogas is obtained from the anaerobic digestion of wet biomass which is a relevant component of most urban residual wastes, as well as industrial food and

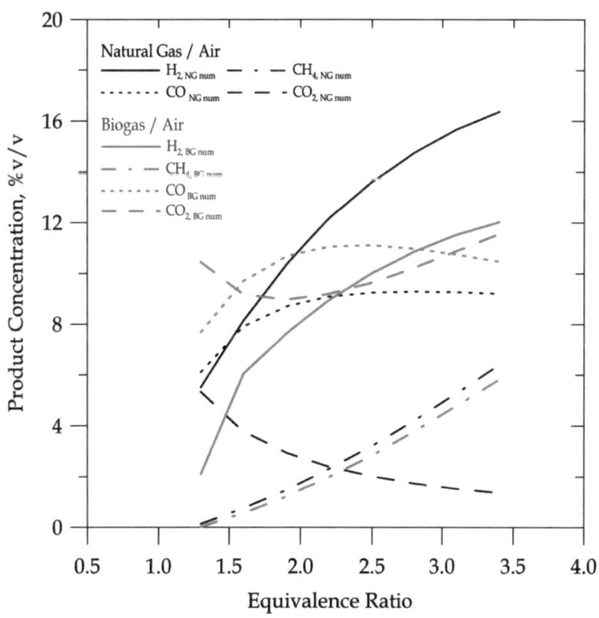

Figure 3.
Composition of chemical products as a function of equivalence ratio for rich and ultrarich mixtures.

agricultural waste. It is known as a gaseous admixture mainly composed of methane (40–65% v/v) and carbon dioxide (35–55% v/v) with traces of hydrogen sulphide (0.1–3.0% v/v), moisture, and other trace contaminants.

This section shows the experimental results of combustion temperatures and syngas production from filtration combustion of synthetic biogas-air mixtures, using different compositions of CH_4 and CO_2 for equivalence ratio of $\varphi = 1.5$ and $\varphi = 2.0$. The experimental temperatures reached presented slight differences (<53 K) between $\varphi = 1.5$ and $\varphi = 2.0$, for the tested biogas-air mixtures (**Figure 4A**). The maximum combustion temperatures experimentally found were 1564 and 1563 K for 55:45 using $\varphi = 1.5$ and $\varphi = 2.0$, respectively. A temperature decrease of 31 K ($\varphi = 1.5$) and 8 K ($\varphi = 2.0$) was observed with an increment of the CO_2 content in the biogas mixtures from 100:0 to 40:60. These peaks in the temperature profile could be associated with the partial oxidation of the CH_4 component of the biogas (exothermic reaction), while the decline in the temperature profile corresponds to the dry reforming of biogas (endothermic reaction). Also, the different behaviours of temperature profile are attributed to the changes of combustion chemistry, CO_2 presence, and filtration velocity, since all other parameters such as heat content (CH_4 flow rates), porosity of inert media, geometry, and dimensions of the reactor were kept constant. The effect of decreasing the filtration velocity while reducing the CO_2 content in the synthetic biogas could be responsible of the similarity between the peak temperatures recorded for all tested conditions, since it is well known that an increasing filtration velocity is responsible of enhancing the diffusion inside the reactor due to larger turbulence inside the pores of the solid matrix, whereas the peak temperature recorded while operating with a biogas composition of 55:45 using $\varphi = 1.5$ and $\varphi = 2.0$ could be attributed to a more intense role of the exothermic reactions in comparison to the decreasing filtration velocity.

Figure 4B illustrates the hydrogen and carbon monoxide yields for the biogas-air mixtures in the inert bed. The maximum hydrogen yields recorded were 17.68 and 15.30% for $\varphi = 1.5$ and $\varphi = 2.0$, respectively, using 100:0 (natural gas-air mixtures). On the other hand, the maximum peak of hydrogen yields using biogas-air mixtures (at 55:45) were 23.34 and 20.40% to $\varphi = 1.5$ and $\varphi = 2.0$, respectively, before gradually declining with the biogas mixture of 40:60.

Similar results have been previously reported by Zeng et al. [36] while operating an inert porous media reactor in a stationary regime with a 50:50 CH_4/CO_2 ratio and a filtration velocity of 25.6 cm/s.

3.4 Steam addition

Variations on the steam to carbon (S/C) ratio from 0.0 to 2.0 were performed under constant values of filtration velocity (34.4 cm/s), an equivalence ratio of $\varphi = 2.0$, and biogas composition (60:40 CH_4/CO_2).

Figure 5A depicts peak operational temperatures, reported as the mean value of maximum temperatures recorded inside the reactor, for several S/C ratios. A decreasing temperature in the reaction zone due to an increasing S/C fraction, which went from 1543 K at baseline conditions to 1501 K at an S/C ratio of 2.0, could be associated to an increased contribution of endothermic reactions in the thermochemical conversion of the mixture. However, since biogas is mainly composed by CH_4 and CO_2, the filtration combustion mode studied could be considered as a tri-reforming process where thermal partial oxidation (TPOX), dry reforming (DRR), and steam reforming (STR) simultaneously interact with CH_4. Thus, this process can be considered as a non-catalytic alternative for biogas valorisation. Regarding the reaction wave propagation rates, **Figure 5B** shows the values computed for each experimental run, which for all experiments propagated in a downstream direction with no relevant variations.

DOI: http://dx.doi.org/10.5772/intechopen.86620

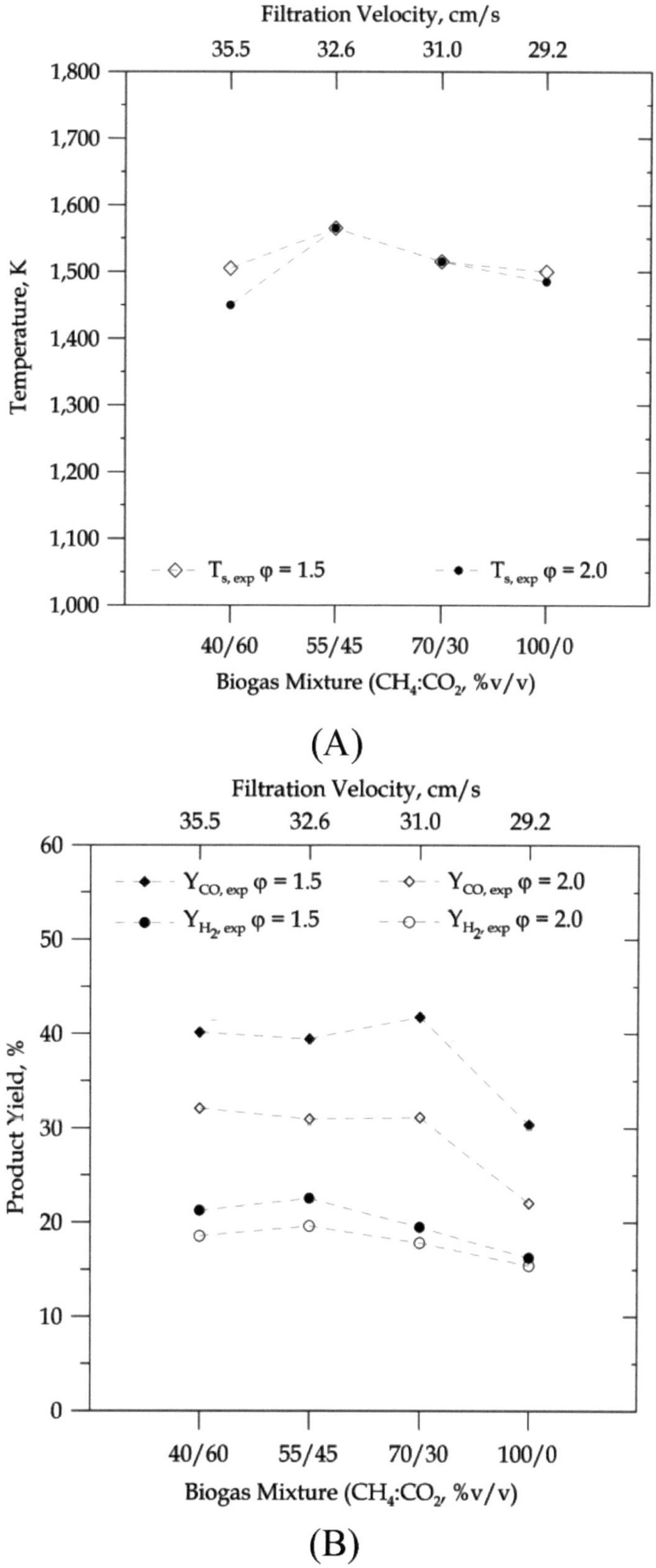

Figure 4.
Combustion temperatures (A) and hydrogen and carbon monoxide yield (B) for equivalence ratio of φ = 1.5 and φ = 2.0, varying biogas composition and filtration velocities.

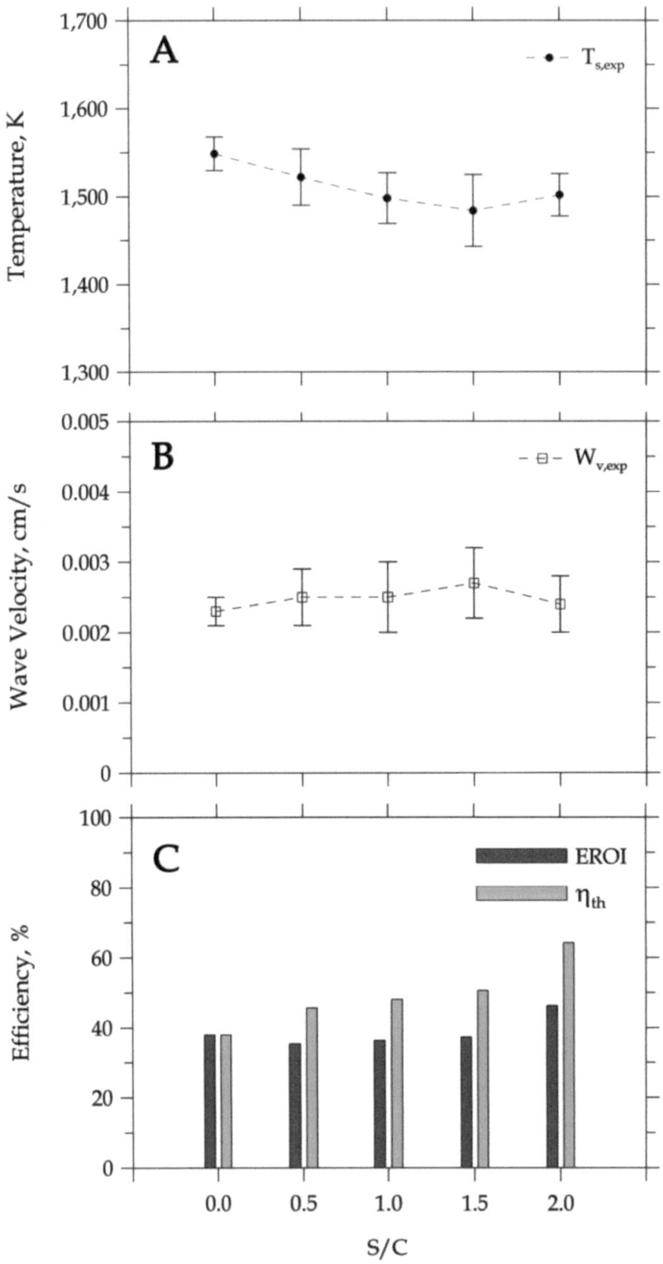

Figure 5.
Combustion temperatures (A), wave propagation rate (B), thermal efficiency, and EROI (C) varying steam to carbon (S/C) ratio. Energy evaluation parameters.

Figure 5C presents the computed values for thermal efficiencies and a simplified energy return on investment (EROI) of the process as a function of the S/C ratio. Overall, both values were positively affected by an increasing presence of steam in the reactants, reaching their maximum values at an S/C fraction of 2.0. In particular, the peak thermal efficiency was accounted as 64.2% which represents an upgrade of 69% when compared to baseline conditions, while the EROI, where both the initial heating value of the biogas and the energy required to supply the steam were considered, reached a maximum of 46.3%, which corresponded to an effective increase of 22% relative to baseline conditions. Therefore, an increasing steam presence in the mixture is considered to favour the effective thermochemical conversion of biogas by means of a non-catalytic filtration combustion reactor.

4. Conclusion

In this chapter numerical and experimental results were presented for filtration combustion of rich biogas-air mixtures, in comparison with methane-air mixtures. Predictions of a numerical model, based on the two-temperature approximation and multistep gas phase combustion mechanism (GRI 3.0), are in good qualitative agreement with experimental data, including combustion temperatures and wave velocities.

Applications for the reforming of biogas fuel with different compositions of methane and carbon dioxide into hydrogen and syngas were presented. Also, some improvement as steam addition to biogas-air mixtures allows higher efficiency for hydrogen and syngas production.

Acknowledgements

The author wishes to acknowledge the financial support by the DGIIP-USM and the technical support of Lorena Espinoza and Nicolás Ripoll.

Author details

Mario Toledo Torres
Department of Mechanical Engineering, Universidad Técnica Federico Santa María, Valparaíso, Chile

*Address all correspondence to: mario.toledo@usm.cl

References

[1] Chen X, Honda K, Zhang ZG. A comprehensive comparison of CH_4-CO_2 reforming activities of NiO/Al_2O_3 catalysts under fixed- and fluidized-bed operations. Applied Catalysis A: General. 2005;**288**:86-97. DOI: 10.1016/j.apcata.2005.04.037

[2] Hou Z, Gao J, Guo J, Liang D, Lou H, Zheng X. Deactivation of Ni catalysts during methane autothermal reforming with CO_2 and O_2 in a fluidized-bed reactor. Journal of Catalysis. 2007;**250**: 331-341. DOI: 10.1016/j.jcat.2007.06.023

[3] Jing Q, Lou H, Fei J, Hou Z, Zheng X. Syngas production from reforming of methane with CO_2 and O_2 over Ni/SrO-SiO_2 catalysts in a fluidized bed reactor. International Journal of Hydrogen Energy. 2004;**29**:1245-1251. DOI: 10.1016/j.ijhydene.2004.01.012

[4] Gao J, Guo J, Liang D, Hou Z, Fei J, Zheng X. Production of syngas via autothermal reforming of methane in a fluidized-bed reactor over the combined CeO_2-ZrO_2/SiO_2 supported Ni catalysts. International Journal of Hydrogen Energy. 2008;**33**:5493-5500. DOI: 10.1016/j.ijhydene.2008.07.040

[5] Moral A, Reyero I, Alfaro C, Bimbela F, Gandía LM. Syngas production by means of biogas catalytic partial oxidation and dry reforming using Rh-based catalysts. Catalysis Today. 2018; **299**:280-288. DOI: 10.1016/j.cattod.2017.03.049

[6] Boullosa-Eiras S, Zhao T, Chen D, Holmen A. Effect of the preparation methods and alumina nanoparticles on the catalytic performance of $Rh/Zr_xCe_{1-x}O_2$-Al_2O_3 in methane partial oxidation. Catalysis Today. 2011;**171**:104-115. DOI: 10.1016/j.cattod.2011.04.021

[7] Kim S, Sung T. Performance and greenhouse gas reduction analysis of biogas-fueled solid-oxide fuel cells for a sewage sludge and food waste treatment facility. Energies. 2018;**11**:600. DOI: 10.3390/en11030600

[8] Zhao X, Li H, Zhang J, Shi L, Zhang D. Design and synthesis of NiCe@m-SiO_2 yolk-shell framework catalysts with improved coke- and sintering-resistance in dry reforming of methane. International Journal of Hydrogen Energy. 2016;**41**:2447-2456. DOI: 10.1016/j.ijhydene.2015.10.111

[9] Artz J, Müller TE, Thenert K, Kleinekorte J, Meys R, Sternberg A, et al. Sustainable conversion of carbon dioxide: An integrated review of catalysis and life cycle assessment. Chemical Reviews. 2018;**118**:434-504. DOI: 10.1021/acs.chemrev.7b00435

[10] Toledo M, Bubnovich V, Saveliev A, Kennedy L. Hydrogen production in ultrarich combustion of hydrocarbon fuels in porous media. International Journal of Hydrogen Energy. 2009;**34**: 1818-1827

[11] Dhamrat RS, Ellzey JL. Numerical and experimental study of the conversion of methane to hydrogen in a porous media reactor. Combustion and Flame. 2006;**144**:698-709

[12] Babkin V, Korzhavin A, Bunev V. Propagation of premixed gaseous explosion flames in porous media. Combustion and Flame. 1991;**87**:182-190

[13] Babkin V. Filtrational combustion of gases. Present state of affairs and prospects. Pure and Applied Chemistry. 1993;**65**:335-344

[14] Vogel BJ, Ellzey JL. Subadiabatic and superadiabatic performance of a two-section porous burner. Combustion Science and Technology. 2005;**177**: 1323-1338

[15] Abdul M, Abdullaha M, Abu Bakarb M. Combustion in porous media and its applications—A comprehensive survey. Journal of Environmental Management. 2009;**90**:2287-2312

[16] Bingue JP, Saveliev AV, Fridman AA, Kennedy LA. Hydrogen production in ultra-rich filtration combustion of methane and hydrogen sulfide. International Journal of Hydrogen Energy. 2002;**27**:643-649

[17] Dobrego KV, Zhdanok SA, Khanevich EI. Analytical and experimental investigation of the transition from low velocity to high-velocity regime of filtration combustion. Experimental Thermal and Fluid Science. 2000;**21**:9-16

[18] Foutko SI, Shabunya SI, Zhdanok SA, Kennedy LA. Superadiabatic combustion wave in a diluted methane-air mixture under filtration in a packed bed. Proceedings of the Combustion Institute. 1996;**25**:1556-1565

[19] Kennedy LA, Bingue JP, Saveliev AV, Fridman AA, Foutko SI. Chemical structures of methane-air filtration combustion waves for fuel-lean and fuel-rich conditions. Proceedings of the Combustion Institute. 2000;**28**: 1431-1438

[20] Drayton MK, Saveliev AV, Kennedy LA, Fridman AA, Li Y-E. Syngas production using superadiabatic combustion of ultra-rich methane-air mixtures. Proceedings of the Combustion Institute. 1998;**27**: 1361-1367

[21] Weinberg FJ, Bartleet TG, Carleton FB, Rimbotti P, Brophy JH, Manning RP. Partial oxidation of fuel rich mixtures in a spouted bed combustor. Combustion and Flame. 1988;**72**:235-239

[22] Ytaya Y, Oyashiki T, Hasatani M. Hydrogen production by methane-rich combustion in a ceramic burner. Journal of Chemical Engineering of Japan. 2002; **35**:46-56

[23] Kennedy LA, Saveliev AV, Fridman AA. Transient filtration combustion. In: Mediterr. Combust. Symp. Vol. 1. 1999. pp. 105-139

[24] Kennedy LA, Saveliev AA, Bingue JP, Fridman AA. Filtration combustion of a methane wave in air for oxygen enriched and oxygen-depleted environments. Proceedings of the Combustion Institute. 2002;**29**:835-841

[25] Howell JR, Hall MJ, Ellzey JL. Combustion of hydrocarbon fuels within porous medium. Progress in Energy and Combustion Science. 1996; **22**:121-145

[26] Gavrilyuk VV, Dmitrienko YM, Zhdanok SA, Minkina VG, Shabunya SI, Yadrevskaya NL, et al. Conversion of methane to hydrogen under superadiabatic filtration combustion. Theoretical Foundations of Chemical Engineering. 2001;**35**:589-596

[27] Gerasev AP. Hybrid autowaves in filtration combustion of gases in a catalytic fixed bed. Combustion, Explosion, and Shock Waves. 2008;**44**: 123-132

[28] Schoegl I, Newcomb SR, Ellzey JL. Ultra-rich combustion in parallel channels to produce hydrogen-rich syngas from propane. International Journal of Hydrogen Energy. 2009;**34**: 5152-5163

[29] Kaviany M. Principles of Heat Transfer in Porous Media. New York: Springer-Verlag; 1991

[30] Henneke MR, Ellzey JL. Modeling of filtration combustion in a packed bed. Combustion and Flame. 1999;**117**: 832-840

[31] Wakao N, Kaguei S. Heat and Mass Transfer in Packed Beds. New York:

Gordon and Breach Science Publications; 1982

[32] Touloukian YS, Powell RW, Ho CY, Nioolsou MC. Thermophysical Properties of Matter. Vol. 10. New York-Washington: IFI/PLENUM; 1973

[33] Smith GP, Golden DM, Frenklach M, Moriarty NW, Eiteneer B, Goldenberg M, et al. Available from: http://www.me.berkeley.edu/gri_mech

[34] Kee RJ, Rupley FM, Miller JA. Report No. SAND89–8009B UC-706; Sandia National Laboratories; 1989

[35] Kee RJ, Grcar JF, Smooke MD, Miller JA. Report No. SAND85-8240 UC-401; Sandia National Laboratories; 1985

[36] Abdul Aziz NIH, Hanafiah MM, Mohamed Ali MY. Sustainable biogas production from agrowaste and effluents—A promising step for small-scale industry income. Renewable Energy. 2019;**132**:363-369. DOI: 10.1016/j.renene.2018.07.149

Section 5

Application of Anaerobic Digestion

Chapter 9

Biofuel: An Environmental Friendly Fuel

Adeola Suhud Shote

Abstract

Various types of biofuels and feedstocks are considered and discussed in terms of their environmental and economic feasibilities. Biofuel is gaining the centre stage as human activities keep rising and the consequent increase in the discharge of lethal emissions is also a subject of concern. The need to cut down greenhouse gas emissions (i.e. CO_2, N_2O, CO, NO, SO_2) is imperative to preserve our natural biodiversity. Biodiesel and bioethanol are the most common, viable alternatives and infinite green fuels that can be used in internal combustion engine. Biodiesel (commonly from waste cooking oil, nonedible vegetable oil, animal fat and tallow) and bioethanol (usually from forestry waste, Lignocellulosic biomass, starchy and sugary vegetable sources, and agricultural residues) are synthesized from straight vegetable feedstocks to bring their characters close to that of the fossil diesel and gasoline. The candidates as green fuels have the potential to significantly reduce the greenhouse gas emissions by as much as 30% from their combustion in internal combustion engine. The various possible methods used for their productions determine the fuel sensitivity to the environment and the energy balance. In general, the energy balances are positive for both fuel substitutes.

Keywords: biofuel, biodiesel, bioethanol, emission, combustion

1. Introduction

The serious concerns of the global climate change, the rising trend of environmental pollution among others have necessitated researchers and industries to develop renewable alternative and cleaner energies across the world. The need for alternative and sustainable energy sources also arises from the global rise in population and the consequent increase in energy demand even as industrialization keeps on expanding. Therefore, the need to drive the world with efficient and eco-friendly energy carriers is paramount. Biofuels are one of the renewable, sustainable sources of energy carriers that can drive the modern day world with the high prospect and potential of reducing greenhouse gas emissions [1–5]. This reduction in this greenhouse gas emission will certainly be in line with Kyoto Protocol. Yet, no fuel system is completely free of environmental concerns [6]. The high dependency and pressure on finite fossil fuels will be shifted due to the global call to look into renewable eco-friendly fuel carriers.

The key sectors that need to be driven by efficient and cleaner fuel substitutes are the aviation, transportation and the manufacturing industries. Over the last two decades, a number of biofuels have been developed. These include bioethanol, biodiesel, biogas, synthetic fuel, hydrogen and so on. Most of these fuels can be blended or used directly in internal combustion engine (ICE) [7, 8]. The cost of production of some these fuels is still relatively on the high side even despite the fact that they have lower environmental consequences when compared with fossil fuel. The various production methodologies could be responsible for the high cost of these biofuels [9]. Syntheses like fermentation, alcoholysis, Saccharification, acidolysis, hydrolysis, esterification, gasification, liquefaction and extraction have been used to produce biofuels.

The combustion of hydrocarbon fuel with O_2 from the atmosphere gives equivalent amount of CO_2 and H_2O. The other discharges are carbon monoxide (CO), nitrogen oxides (NO, N_2O) and nitrogen compounds (NH_3 and HCN), sulfur gases (SO_2, CS_2, OCS), compounds of halogens and carbons (CH_3Br and CHCl). Combustion of these finite fossil fuels and other biomass led to the global alteration in the atmosphere with respect to huge emissions discharges from these two main contributors [10]. Combustion from automobile, stationary sources are largely responsible for most of the greenhouse gas emissions to date (through CO_2, stratospheric O_3, and soot) and also some opposing effects (through SO_2). It has had minimal effect on stratospheric O_3 (through CH_3Cl, CH_3Br, CH4) but has likely influenced the stratospheric oxidant levels (through CO, NO_x, NMHC), specifically the Northern Hemisphere [10]. Even though energy is key to drive the daily and economic activities but it should not be to the detriment of humanity.

Bioethanol, biogas and biodiesel are the most widely used biofuels. Bioethanol production is very high in Brazil and USA due to large volume of production of sugar cane and corn respectively. Research is also directed to the use of cellulose to produce ethanol [11]. Cellulose is converted to sugar and thereafter to ethanol. However, biodiesel is common in the Scandinavian countries and Germany in Europe. Biodiesel is usually blended using about 5–20% in these countries [12]. Germany is one of the top countries known for the production of biodiesel. Biogas on the other hand is usually produced from animal waste. The leading countries in the production of biogas in Europe are Germany and Great Britain. As the production of biofuel is gaining interest, bio-refineries are being cited in advanced countries like the USA, Germany to produce biofuels and other associated products.

Gasification of biomass to produce methanol, ethanol, dimethylether, syn-diesel could also result in the production of hydrogen and methane which may be used in vehicles. Largely, all these conversion processes are still relatively high in cost.

1.1 Biofuel as an alternative fuel

Researches are going on in the use of sustainable alternative fuels for automobile and stationery machines. The most prominent of those technologies being under investigation are the use of electric and hybrid automobiles, compressed natural gas (CNG), dimethylether, hydrogen, liquid biofuel, liquefied petroleum gas (LPG) and liquefied natural gas (LNG) among others. The centre of discussion here will focus on two prominent liquid fuels which are biodiesel and bioethanol. Their environmental impact and their sensitivity analysis will be looked into in detail.

These fuels are sourced mainly from vegetable feedstock chain supplies. These may include edible and nonedible vegetable sources that are sustainable [1].

2. Bioethanol

The feedstocks include sorghum, sugar beet, wheat, cassava and so on. Bioethanol can be used in straight or even blended form with premium motor spirit

(PMS) in spark ignition engine (SIE). Engine modification is not necessary for blends between 5 and 20%. The blend is often represented as E(percentage)G, that is E15G. The middle term '15' represent the percentage of the blend. However, higher blends may require engine modification. This adds to the final cost of the use of bioethanol in SIE. Besides, many advantages are associated with the candidate's use in SIE. The blends help in engine lubrication, thereby reducing the wear rate and reducing the engine temperature. It also has the advantage of cutting down the greenhouse gas emissions [13].

2.1 Pre-treatment for bioethanol

The production and quality of bioethanol produced is significantly boosted by the pretreatment procedures. Various pretreatment methods are presented in [14, 15] for lignocellulosic substrates and the merits of employing a pretreatment procedure. Pretreatment is necessary for the substrate feedstock because it gives direct yield of fermentable sugar ready for hydrolysis and heating. It also inhibits degradation and retards the activities of inhibitors for the final conversion to ethanol [15]. It gives a positive energy balance at the long run as the pretreatment process reduces the production cost, wastage of materials and time before the final fermentation process [15, 16]. The pretreatment processes involve the reduction of the size of the feedstock matrix-fraction (some time through the use of enzyme called amylase or through the mechanical means like milling). Hydrolysis and heating can then follow so that the enzyme can easily acts on the pretreated substrate [15].

Microorganisms like fungi can also be employed for the pretreatment of feedstocks particularly lignocellulosic materials. This involves breaking of the lignin structure. This pretreatment method is usually used on a laboratory scale because the process is slow and does not require much energy input as chemical processes are not involved [17].

The chemical pretreatment (alkaline/acid) is desirable for large scale industrial production due to the availability and affordability of chemical which are not affected by ambient changes unlike biological pretreatment agents [14]. Diluted acid pretreatment is usually preferred due to the draw backs of the use of concentrated acid pretreatment like corrosion of production line components. Within a short period of time and at about 160°C, sugar monomers are formed. The common agents used for acid pretreatment include nitric acid (HNO_3), hydrochloric acid (HCl), sulfuric acid (H_2SO_4), organic acid (lactic/acetic acids) and that of the alkaline are sodium hydroxide ($NaOH$), calcium hydroxide ($Ca(OH)_2$) ammonium hydroxide (NH_4OH) [14]. The various pretreatment methods and effects are summarized in **Table 1**.

2.2 Synthesis of bioethanol

The traditional method involves liquefaction and saccharification of starch. Saccharification involves the hydrolysis of cellulose or starch (polysaccharides) to simple monosaccharides. Carbohydrates (such as sucrose, maltose) are broken down to give simple sugar (like glucose, fructose, galactose).

The process is usually catalyzed using biological method (an enzyme) or chemical method (acid/base) [15, 18]. The saccharification procedure is very important because it determines the time, quality and quantity of the final product which is the bioethanol. Crystalline nature of cellulose fiber, lignin and hemicellulose content, porosity of lignocellulosic substrate affect the hydrolysis process [19]. Some drawbacks like increase in process cost from the chemical obtained, acid recovery, corrosion of production line components are associated with the use of chemical

Pretreatment method	Effects	Pros	Cons
Diluted acid	Hydrolyses hemicelluloses; Alters the structure of lignin Reduces cellulose more amenable for additional enzymatic treatment	Less corrosion effect than the conc acid Low formation of inhibitors	Generation of degradation products due to high temperature Low sugar concentration exit stream
Concentrated acid	Hydrolyses both hemicelluloses and cellulose	High glucose yield Operational cost reduction due to moderate operating temperature Low formation of degradable products No enzyme are required	Acid recovery is mandatory Equipment corrosion Generation of inhibitory compounds
Alkali	Removes lignin and hemicelluloses Increases accessible surface area	High lignin removal High digestibility	Long residence time Irrecoverable salt formation
Biological	Degrades lignin and hemicelluloses	Low energy consumption	Low hydrolysis rate
Mechanical	Reduces cellulose crystallinity	No formation of inhibitors	High power and energy consumption
Ammonia fiber explosion	Increases accessible surface area Slightly removes lignin and hemicelluloses to an extent	Low formation of inhibitors	Not efficient for biomass with high lignin content High cost due to large amount of ammonia
Ammonia recycled percolation	Removes lignin	Highly selective delignification	High energy consumption
Ionic liquids	Reduces cellulose crystallinity Removes lignin	High digestibility Green solvent	Large-scale application still under investigation
Ozonolysis	Reduces lignin transformation	No formation of inhibitors Mild operational conditions	High cost due large amount of ozone required
Steam explosion	Causes lignin transformation Causes hemicellulose solubility	Cost effective High yield of glucose and hemicellulose	Generation of inhibitory compounds Partial hemicellulose degradation Incomplete disruption of the lignin carbohydrate matrix
Supercritical fluid technology	Increase accessible surface area	Cost effective No formation of inhibitor	Does not affect lignin and hemicelluloses Very high pressure requirements
Wet oxidation	Removes lignin	Low formation of inhibitors	High cost of oxygen and alkaline catalyst
Organosolv	Hydrolyses linin and hemicelluloses	Pure lignin recovery High digestibility	Solvents need to be drained and recycled High cost

Table 1.
Pretreatment methods for lignocellulosic feedstocks in bioethanol synthesis [14].

catalysis [20]. However, the use of dilute acid is often preferred as it contributes to less impact on the environment. The enzymatic approach is mostly carried out in an orbital shaker (120–150 rpm) at about 40–50°C for 120 h [21]. β-Glucosidase, *Cellulase* are commonly used for the hydrolysis of complex starch [22]. Thereafter, the fermentation of sugar extracted from starch plant, followed by distillation. Enzyme, bacteria or yeast are used during the fermentation process to accelerate the formation of the product in an anaerobic environment. The simple sugar is converted to ethanol and carbon dioxide. Some time, saccharification and fermentation are considered simultaneously due to low cost, reduced process time, prevention of cellulose inhibition, high yield of ethanol [22]. Most of the world's ethanol is produced in Brazil and USA. Corn is used to synthesize bioethanol in the USA while sugar cane is used substantially in Brazil [13]. However, in Spain, bioethanol is produced from barley and in France it is obtained from beet. Agricultural biomass or residue from waste paper or corn stalks can also be used to obtain bioethanol [13]. Other agricultural produce that can be used to generate ethanol are wheat, potato, maize cassava etc.

Ethyl tertiary butyl ether (ETBE) is also worthy of mentioning here since it can also be obtained from some of the agricultural produce like barley, wheat which can be blended with PMS. They are usually produced in the refinery. Lignocellulose residue is gaining more research interest due to the reduction in the cost of production [14]. The lignocellulose contains cellulose, hemicelluloses and lignin. Hydrolysis are used to fractionalize some agricultural produce like maize so that complete conversion to ethanol and carbon dioxide can be obtained easily and efficiently with reduced wastage.

2.3 The ethanol conversion process

The schematic process of production chain of bioethanol is depicted in **Figure 1**. The process starts with milling of the feedstock for easy extraction of the starch. Yeast is added to make the extraction easy. The next stage is hydrolysis. Enzyme is added to obtain single sugar (glucose) and starch before fermentation by bacteria or microorganisms. Thereafter, distillation is followed to remove large volume of water in the product and then dehydration to further remove the water content so that the outcome concentration will increase. There are various pretreatment methods that can be used to obtain high quality ethanol [14, 23]. **Figure 2** presents the summary of the pretreatment and conversion methods while **Table 1** presents the pretreatment methods and the various merits and drawbacks from the use of

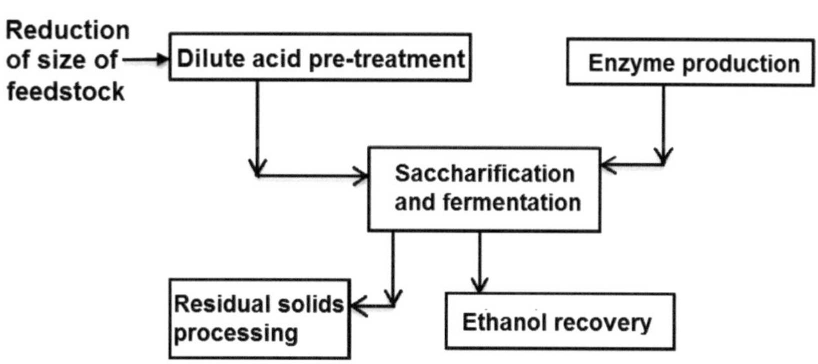

Figure 1.
Bioethanol conversion scheme [29].

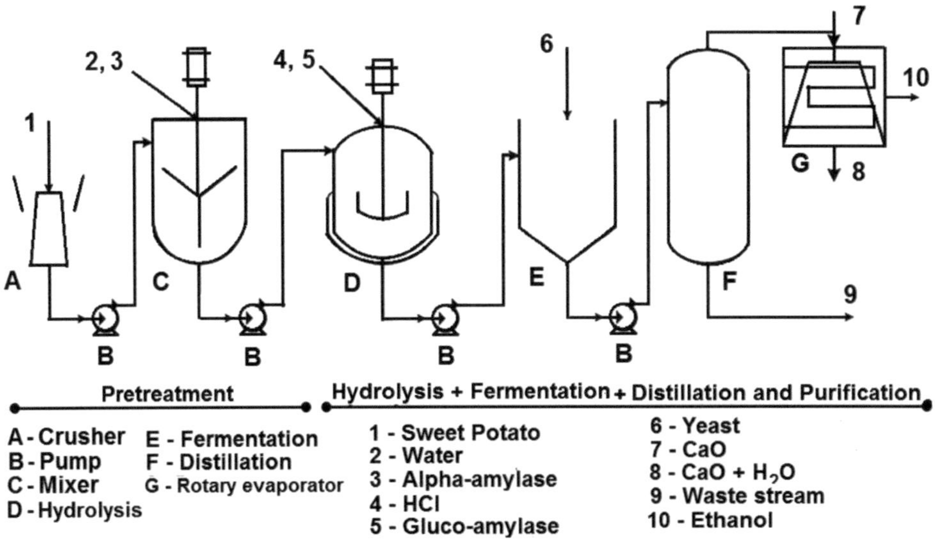

Figure 2.
Pretreatment and bioethanol conversion processes [27].

respective pretreatment method. In more recent time, response surface methodology (RSM) and artificial neural network (ANN) have been used to optimize the production of bioethanol [24]. ANN is used to perform complex cognitive procedurals which mimic the biological function of human brain. ANN genetic algorithm is used to simulate the biological processes in the ethanol production so that the major parameters can be varied to obtain a much desired output.

2.4 Impact of bioethanol and its sensitivity on the environment

This is related to the emission pattern compared with the fossil premium motor spirit (FPMS) and the energy balance for the synthesis of the ethanol. The energy balance involves the input feedstock and the by-product. Bioethanol has a huge potential to reduce the hazardous gas emissions due to its vegetable source. It helps in the complete combustion process because it is oxygenated thereby reducing the emissions by as much as 30%. Bioethanol from lignocellulosic biomass resource has huge strategic potential in cutting down of greenhouse gas emissions into the environment and reducing the consumption of crude fuel [21]. It can be blended with PMS due to its characteristics such as low cetane number, high octane number, high heat of vaporization that are close to that of the fossil gasoline [25]. The use of this large lignocellulosic biomass has the potential to reduce transportation cost which will contribute tremendously to positive energy balance [26]. Ethanol production from lignocellulose also has advantages over first generation biofuel in that it makes use of low cost biomass and employs a production process that is environmentally friendly [27].

Various methods of biofuel feedstocks, production loading methodologies and the environmental and economic viability of biofuel production are issues that are attracting and promoting the use of biofuels. However, biofuels combustion in ICE were found to have the prospect of reducing the greenhouse gas emissions [28] thereby lowering the average mean temperature of the atmosphere in the long run. This is in line with Kyoto protocol. The global pressure on the finite premium motor spirit (PMS) and automotive gas oil (AGO) will reduce significantly by using biofuel in straight or blended forms. Biofuels also have the merits of renewability and sustainability. The energy balance of most biofuels is positive as production is maximized.

3. Biodiesel

Biodiesel is currently gaining more and more interest as the desire to use eco-friendly fuel substitute keeps increasing because of the increase in the fluctuations in the crude-fuel prices internationally and the environmental consequences of fossil fuel. There are numerous ways of producing biodiesel from different feedstocks and different catalysts. Some of the feedstock include but not limited to grapeseed, camelina, lupin, linseed, rapeseed, sunflower, peanut, palm oil, palm kernel oil, poppyseed, olive, chestnut, karanja, pongamia, soybeans, canola, corn, crambe, jatropha, cottonseed and so on. Animal fat can also be used to synthesize biodiesel. The common methods of production are transesterification (alkaline catalyzed), the use of enzyme (e.g. lipase) catalyst, supercritical method of production (under high temperature and relatively high pressure) and the use of acid catalyst [30]. However, the commonly used method for the production of biodiesel is the transesterification method because of its simplicity and ease of handling.

Vegetable oils normally contain fatty acids which makes the properties (like viscosity, density, flash point, pour point cloud point, cetane number) to be high in value and mostly unsuitable to be used in internal combustion engine (ICE) without engine modification. So, the essence of the conversion to biodiesel is to remove or lower the effect of the fatty acid composition in **Table 2** [30]. The chemical structures of common fatty acids are also presented in **Table 3**. Some oils that are having lower fatty acid content can be used directly or in blended form as biodiesel fuel in compression ignition engine (CIE) which may require modification. This makes the whole chain of production more costly.

'xx' indicates number of carbons, and 'y' number of double bonds in the fatty acid chain.

The production of biodiesel starts with oil extraction (usually with hexane). The high fatty acid oil containing triglycerides reacts with alcohol (usually methanol or ethanol) to produce esters and glycerol in the presence of a catalyst (KOH or NaOH). The whole processes are reversible processes and are in three steps as shown in Eqs. (1)–(3) [16]. Excess alcohol is desirable to accelerate the reaction towards the products side. The kinetics parameter k_x are in Noureddini and Zhu [32].

Transesterification of rapeseed oil in the supercritical methanol method shows that at temperature of 239°C and a pressure of 8.09 MPa, glycerin and methyl esters are produced as the principal products [31].

The main product is biodiesel and the by product is glycerol which are mainly used in the cosmetics and pharmaceutical industries. The basic conversion scheme of the vegetable oil is shown in **Figure 3**.

Fatty acid	Lauric	Myristic	Palmitic	Stearic	Oleic	Linoleic	Linolenic
Soybean	0.1	0.1	10.2	3.7	22.8	53.7	8.6
Palm	0.1	1.0	42.8	4.5	40.5	10.1	0.2
Cotton seed	0.1	0.7	20.1	2.6	19.2	55.2	0.6
Tallow	0.1	2.8	23.3	19.4	42.4	2.9	0.9
Coconut	46.5	19.2	9.8	3.0	6.9	2.2	0.0
Lard	0.1	1.4	23.6	14.2	44.2	10.7	0.4

Table 2.
Fatty acid in vegetable oil [30].

$$\text{Triglyceride (TG)} + \text{R}'\text{OH} \underset{k_4}{\overset{k_1}{\leftrightarrow}} \text{Diglyceride (DG)} + \text{R}'\text{COOR1} \tag{1}$$

$$\text{Diglyceride (DG)} + \text{R}'\text{OH} \underset{k_5}{\overset{k_2}{\leftrightarrow}} \text{Monoglyceride (MG)} + \text{R}'COOR2 \tag{2}$$

$$\text{Monoglyceride (MG)} + \text{R}'\text{OH} \underset{k_6}{\overset{k_3}{\leftrightarrow}} \text{Glycerol (GL)} + \text{R}'COOR3 \tag{3}$$

3.1 Impact of biodiesel and its sensitivity on the environment

This centers on the emissions comparison of the finite fossil diesel with biodiesel. Shote et al. [5] conducted a research on the emission patterns of the biodiesel blends compared with petroleum diesel. One hundred percent biodiesel was found to have lowest impact on the environment in terms of the hazardous emissions (CO, NO, NO_2, NO_x). Besides, the result also shows gradual reduction of the emissions pattern as the concentration of PKO-based biodiesel increases in the blend.

Fatty acid	Chemical name of fatty acids	Structure (xx:y)	Formula
Lauric	Dodecanoic	12:1	$C_{12}H_{24}O_2$
Myristic	Tetradecanoic	14:1	$C_{14}H_{28}O_2$
Palmitic	Hexadecanoic	16:0	$C_{16}H_{32}O_2$
Stearic	Octadecanoic	18:0	$C_{18}H_{36}O_2$
Arachidic	Eicosanoic	20:0	$C_{20}H_{40}O_2$
Behenic	Docosanoic	22:0	$C_{22}H_{44}O_2$
Lignoceric	Tetracosanoic	24:0	$C_{24}H_{48}O_2$
Oleic	cis-9-Octadecenoic	18:1	$C_{18}H_{34}O_2$
Linoleic	cis-9,cis-12-Octadecadienoic	18:2	$C_{18}H_{32}O_2$
Linolenic	cis-9,cis-12,cis-15-Octadecatrienoic	18:3	$C_{18}H_{30}O_2$
Erucle	cis-13-Docosenoic	22:1	$C_{32}H_{42}O_2$

Table 3.
Chemical structure of common fatty acids [31].

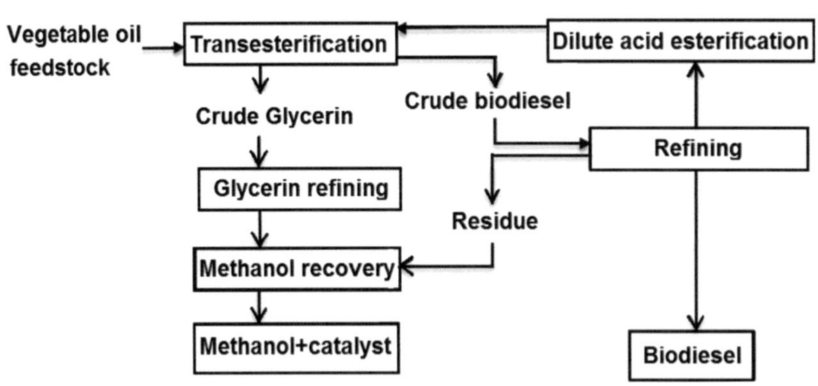

Figure 3.
Biodiesel conversion scheme [16].

In general, the energy balance of most biodiesels is positive depending on the source of the vegetable oil, production, extraction and esterification. Zhang et al. [9] found out that the plant capacity, feedstock of oil and the method used to synthesize biodiesel are the most significant factors determining the final cost of biodiesel. The use of glycerol in the cosmetics and pharmaceutical industries will boost the energy balance of the biodiesel significantly. Since biodiesel is gaining more ground in terms of the usage, then the energy balance tends to be more positive. Biodiesel are found to reduce greenhouse gas emissions [5, 7]. Biodiesels also have the advantage of lubricating and bringing down the temperature of ICE. The generic NO_x emissions are not significantly affected as the concentration of the blends increase in the fuel mixture [5]. NO_x formation is governed by Zeldovich mechanism.

Acid catalyst (tetraoxosulphate VI acid) is also used instead of alkali catalyst (NaOH or KOH) to lower the activation energy for quick formation of esters and glycerin. However, the reaction takes longer time (about 2 days) to complete. The molar ratio is kept at 30:1 and the temperature range is between 60 and 80°C. The triglyceride is usually used with one mole of sulfuric acid to give 90% conversion to ester and glycerin in about 2 days [16].

Other methods involve the use of enzyme to fast track the rate of chemical reaction and the formation of esters. This method is very expensive because of the cost of enzyme which invariably affects the energy balance of the entire chain of production processes.

The summary of the common methods used for the production of biodiesel are presented in **Table 4**.

3.2 The summary of technologies at developmental stage dealing with biofuel production

a. *Pyrolysis of oil*: lignocellulosic biomass is usually used to synthesize bioethanol. The process normally starts with pretreatment procedures. However, any of the biomass substrates can be used to produce biodiesel. Research is till on going in the development of reactor for accelerated pyrolysis processes. A lot of investment and researches are still needed in this area so that the process and the end results will be economically reliable [33].

Variable	Alkali catalyst	Lipase catalysis	Supercritical alcohol	Acid catalysis
Reaction temperature (°C)	60–70	30–40	239–385	55–80
Free fatty acid in vegetable oil	Saponified products	Methyl esters	Esters	Esters
Water in vegetable oil	Interference with reaction	No influence	—	Interference with reaction
Yield of methyl esters	Normal	Higher	Good	Normal
Recovery of glycerol	Difficult	Easy	—	Difficult
Purification of methyl esters	Repeated washing	None	—	Repeated washing
Production cost of catalyst	Cheap	Relatively expensive	Medium	Cheap

Table 4.
Different techniques for ester production [16].

b. *Hydrothermal upgrading process*: this process involves hydrolysis of biomass at high pressure and moderate temperature to produce bio-crude. It is also at the development sage just like pyrolysis process.

c. *Dimethylether*: this is produced from gasification of biomass. It could also be obtained from natural gas. It is one of the conventional diesel fuel substitutes that is capable of cutting down NO_x emissions from its combustion in CIE. It is commonly synthesized from methanol.

d. *Fischer-Tropsch*: synthetic gas is produced from fossil fuels. Research is still on going to produce synthetic gas from biomass feedstock. In the past, the primary feedstock for its production is fossil fuel. The origin of this synthetic fuel can be traced to Germany. Production of synthetic fuel through pyrolysis of biomass is still under investigation.

e. *Synthetic fuel*: this fuel is synthesized by recycling organic waste. A lot of work is going on in this area to optimize the processes. It can be blended with conventional fuel.

Acknowledgements

The author gratefully acknowledged Obafemi Awolowo University, Ile-Ife for her support.

Author details

Adeola Suhud Shote
University of Pretoria, Hatfield, South Africa

*Address all correspondence to: adeolashoteyahoo.ca

References

[1] Balat M, Balat H. Progress in biodiesel processing. Applied Energy. 2010;**87**:1815-1835. DOI: 10.1016/j.apenergy.2010.01.012

[2] Fergione J, Hill J, Tilman D, Polasky S, Hawthorne P. Land clearing and the biofuel carbon debt. Science. 2008;**319**: 1235-1238. DOI: 10.1126/science.1152747. Epub 2008 Feb 7

[3] Scharlemann JPW, Laurance WF. How green are biofuels? Science. 2008; **319**:43-44. DOI: 10.1126/science.1153103

[4] Searchinger T, Heimlich R, Houghton RA, Dong F, Elobeid A, Fabiosa J. Use of U.S. croplands for biofuels increases greenhouse gases through emissions from land use change. Science. 2008;**319**:1238-1240. DOI: 10.1126/science.1151861. Epub 2008 Feb 7

[5] Shote AS, Betiku E, Asere AA. Characteristics of CO and NOx emissions from combustion of transmethylated palm kernel oil-based biodiesel blends in a compression ignition engine. Journal of King Saud University—Engineering Sciences. 2018;1-6. DOI: 10.1016/j.jksues.2018.02.005. In Press

[6] Hoekman SK. Biofuels in the U.S.—Challenges and opportunities. Renewable Energy Journal. 2009;**34**:14-22. DOI: 10.1016/j.renene.2008.04.030

[7] Rahman MM, Pourkhesalian AM, Jahirul MI, Stevanovic S, Pham PX, Wang H, et al. Particle emissions from biodiesels with different physical properties and chemical composition. Fuel. 2014;**134**:201-208. DOI: /10.1016/j.fuel.2014.05.053

[8] Cooney CP, Worm JJ, Naber JD. Combustion characterisation in an internal combustion engine with ethanol-gasoline blended fuels varying compression ratios and ignition timing. Energy & Fuels. 2009;**23**(5):2319-2324. DOI: 10.1021/ef800899r

[9] Zhang Y, Dube MA, McLean DD, Kates M. Biodiesel production from waste cooking oil: 2. Economic assessment and sensitivity analysis. Bioresource Technology. 2003;**90**: 229-240. DOI: 10.1016/so960-8524(03)00150-0

[10] Prather MJ, Logan JA. Twenty-Fifth Symposium (International) on Combustion. Pittsburgh: The Combustion Institute; 1994. p. 1513. DOI: /10.1016/S0082-0784(06)80796-4

[11] Schifter I, Diaz L, Rodriguez R, Gomez JP, Gonzalez U. Combustion and emissions behaviour for ethernol—Gasoline blends in a single cylinder engine. Fuel. 2011;**90**(12):3586-3592. DOI: /10.1016/j.fuel.2011.01.034

[12] Shote AS, Betiku E, Asere AA. Biodiesel production by transmethylation of Nigerian palm kernel oil. Ife Journal of Technology. 2009;**18**(2):1-4. DOI: http://ijt.oauife.edu.ng

[13] Hanif M, Mahlia TIM, Aditiya HB, Abu Bakar MS. Energy and environmental assessments of bioehtanol production from Sir Kanji 1 cassava in Malaysia. Biofuel Research Journal. 2017;**13**:537-544. DOI: 10.18331/BRJ2017.4.1.3

[14] Refaat AA. 5.13—Biofuels from waste materials. In: Sayigh A, editor. Comprehensive Renewable Energy. 1st ed. Oxford: Elsevier; 2012. p. 217–261 DOI: /Letcher/978-0-08-087872-0. Ch5

[15] Aditiya HB, Mahlia TMI, Chong WT, Hadi N, Sebayang AH. Second

generation bioethanol production: A critical review. Renewable and Sustainable Energy Reviews. 2016;**66**: 631-653. DOI: 10.1016/j.rser.2016.07.015

[16] Gupta A, Verma JP. Sustainable bio-ethanol production from agro-residues: A review. Renewable and Sustainable Energy Reviews. 2014;**2015**:550-567. DOI: 10.1016/j.rser.2014.08.032

[17] Sun Y, Cheng J. Hydrolysis of lignocellulosic material for ethanol production: A review. Bioresource Technology. 2002;**98**:673-686. DOI: 10.1016/s0960-8524(01)00212-7

[18] Prado JM, Lachos-perez D, Forster-carneiro T, Rostagno MA. Food and bioproducts processing sub- and supercritical water hydrolysis of agricultural and food industry residues for the production of fermentable sugars: A review. Food and Bioproducts Processing. 2015;**98**:95-123. DOI: 10.1016/j.fbp.2015.11.004

[19] Karimi K, Emtiazi G, Taherzadeh MJ. Ethanol production from dilute acid pretreated rice straw by simultaneous saccharification and fermentation with Mucor indicus, *Rhizopus oryzae*, and *Saccharomyces cerevisiae*. Enzyme and Microbial Technology. 2006;**40**: 138-144. DOI: 10.1016/j.enzmictec.2005.10.046

[20] Noomtim P, Cheirsilp B. Production of butanol from palm empty fruit bunches hydrolyzate by *Clostridium acetobutylicum*. Energy Procedia. 2011;**9**: 140-146. DOI: 10.1016/j.egypro.2011.09.015

[21] Derman E, Abdulla R, Marbawi H, Sabullah MK. Oil palm empty fruit bunches as a promising feedstock for bioethanol production in Malaysia. Renewable Energy. 2018;**129**:285-298. DOI: 10.1016/j.renene.2018.06.003

[22] Raman JK, Gnansounou E. Ethanol and lignin production from Brazilian empty fruit bunch biomass. Bioresource Technology. 2014;**172**:241-248. DOI: 10.1016/j.biortech.2014.09.043

[23] Sathendraa ER, Baskarb G, Praveenkumara R, Gnansounou E. Bioethanol production from palm wood using *Trichoderma reesei* and *Kluveromyces marxianus*. Bioresource Technology. 2019;**271**:345-352. DOI: 10.1016/j.biortech.2018.09.134

[24] Baskar G, Selvakumari AE, Aiswarya R. Biodiesel production from castor oil using heterogeneous Ni doped ZnO nanocatalyst. Bioresource Technology. 2018;**250**:793-798. DOI: 10.1016/j.biotech.2017.12.010

[25] Tan KT, Lee KT, Mohamed AR. Role of energy policy in renewable energy accomplishment: The case of second-generation bioethanol. Energy Policy. 2008;**36**:3360-3365. DOI: 10.1016/j.enpol.2008.05.016

[26] Silva JOV, Almeida MF, Alvim-Ferraz MC, Dias JM. Integrated production of biodiesel and bioethanol from sweet potato. Renewable Energy. 2018;**124**:114-120. DOI: 10.1016/j.renene.2017.07.052

[27] Boonchuaya P, Techapunb C, Leksawasdib N, Seesuriyachanb P, Hanmoungjaib P, Watanabec M, et al. An integrated process for xylooligosaccharide and bioethanol production from corncob. Bioresource Technology. 2018;**256**:399-407. DOI: 10.1016/j.biortech.2018.02.004

[28] Lee S, Teramoto Y, Endo T. Enzymatic saccharification of woody biomass micro/nanofibrillated by continuous extrusion process I—Effect of additives with cellulose affinity. Bioresource Technology. 2009;**100**: 275-279. DOI: 10.1016/j.biotech.2008.05.051

[29] Lin Y, Tanaka S. Ethanol fermentation from biomass resources:

Current state and prospects. Applied Microbiology Biotechnology. 2006;**69**: 627-642. DOI: 10.1007/s00253-005-0229-x

[30] Marchetti JM, Miguel VU, Errazu AF. Possible methods for biodiesel production. Renewable and Sustainable Energy Reviews. 2007;**11**:1300-1311. DOI: 10.1016/j.rser.2005.08.006

[31] Barnwal BK, Sharma MP. Prospects of biodiesel production from vegetable oils in India. Renewable and Sustainable Energy Reviews. 2005;**9**:363-378. DOI: 10.1016/j.rser.2004.05.007

[32] Noureddini H, Zhu D. Kinetics of transesterification of soybean oil. Journal of American Oil Chemist's Society. 1997;**74**:1457-1463. DOI: 10.1007/s11746-997-0254-2

[33] Chacon FAT. Techno-economic assessment of biofuel production in the European Union (thesis). Germany: Faculty of Business Admin, Technische University Freiberg; 2004. DOI: http://www.globalbioenergy.org

Chapter 10

Experimental Study of CO_2 Plasticization in Polysulfone Membrane for Biogas Processing

Serene Sow Mun Lock, Kok Keong Lau, Azmi Mohd Shariff, Yin Fong Yeong and Norwahyu Jusoh

Abstract

Polymeric membranes have emerged for biogas processing to remove CO_2 from CH_4. Nonetheless, it is also acknowledged that polymeric membranes have the tendency to sorb highly condensable CO_2, which consequently swells the polymeric matrix, typically at operating condition higher than the plasticization pressure. The swelling increases void spaces for transport of gas penetrants, which results in an increment in permeability of all gas components at the cost of substantial decrease in membrane selectivity. Despite observations of the end results of plasticization, it is found that many transport property studies include only permeability measurements near ambient conditions. Complementary information on the individual contributions of the sorption and diffusion coefficients to the overall performance typically at non-ambient operating conditions is rarely reported. Therefore, in present study, experimental study has been conducted to fabricate polysulfone (PSF) film. Validity of the developed polysulfone membrane has been verified through characterization and validated with gas transport behavior of published results. Subsequently, transport properties of CO_2 though the PSF membrane at varying operating temperatures has been elucidated. The dual mode sorption and partial immobilization models have been employed to quantify the gas transport properties of noncondensable CH_4 and condensable CO_2 through PSF membrane.

Keywords: membrane, plasticization, solubility, diffusivity, permeability

1. Introduction

The ever-growing worldwide energy demand has directed the attention of government agencies and energy companies towards uncovering renewable energy over recent years as an alternative to achieve sustainable global energy policy [1]. The effort is done to circumvent the volatility of fuel price in the petrochemical market while meeting expanding user demand [2]. Biogas produced from microbial digestion of waste is found to contain high concentration of methane (CH_4), which can be utilized for combustion process to circumvent usage of fossil fuels while meeting energy demand. Nonetheless, biogas also contains a huge amount of side products, typically carbon dioxide (CO_2), whereby the amount can reach as high as 50% [1].

It is highly desirable to remove the CO_2 contaminants from CH_4 since the unequivocal symptoms of climate change have urged continuous pressure on oil and gas companies to adopt practices that reduce carbon footprint to mitigate the effect of greenhouse gases global warming [3]. In addition to minimization of the environment pollution, the undesired CO_2 must be removed in order to increase the heating value of biogas since the abundant impurities constitute to no heating value [4]. The removal of CO_2 in the biogas also prevents corrosion of pipelines and process equipments that are of great importance to curb gas leakage along the transportation process since the leakage can contribute to public hazards [5]. It has been proposed that the produced biogas requires processing to contain a minimal of 95% CH_4 in order to be economically viable [1].

Polymeric membrane applied in gas separation is an alternative that has gained attention in industrial scale application in comparison to conventional technologies (e.g. distillation and absorption) over recent years. The advantages associated to polymeric membrane include taking up a considerably confined space, merely involves physical separation that is free from chemical reaction for consideration of process safety, lower energy consumption and smaller operating cost requirement [6, 7]. Polymeric membrane has been utilized exceptionally in application of CO_2 removal from CH_4 for processing of biogas. However, a problem that hinders further expansion of the usage of polymeric membranes in such application has emerged due to CO_2 induced plasticization.

During CO_2 plasticization, sorption of condensable gas penetrants in the membrane matrix interacts with functional group of the pristine polymeric chain. The interaction contributes to ease of mobility of polymeric chains, which consequently increases the void channels in the membrane [8]. Schematic representation of the plasticization phenomena that increases free volume of the polymeric glassy membrane is provided in **Figure 1** [9].

As a result, plasticization increases void channels that form passage for gas permeation of all gas species [10]. Nonetheless, when empty spaces increase, the sieving capability of the polymeric membrane also reduces simultaneously. This causes reduction in the membrane selectivity $\left(\propto_{A/B} = P_A/P_B\right)$ as an ultimate result. Therefore, it is vital to understand CO_2 plasticization in polymeric membranes since it is highly possible to cause undesirable product lost that decreases profitability of the biogas processing plant.

The observation of CO_2 plasticization in glassy polymeric membranes has been well addressed with a long history. Wessling et al. conducted experiments comparing the kinetics of mass uptake (sorption) and the volume increase (dilation) due to sorption to give a deeper understanding of the plasticizing effect of CO_2 in commercial 6FDA membrane [11]. Houde et al. employed the wide angle X-ray diffraction (WAXD) to investigate the mechanism of plasticization in various glassy polymers [12]. Bos et al. reported CO_2 plasticization phenomena, which includes

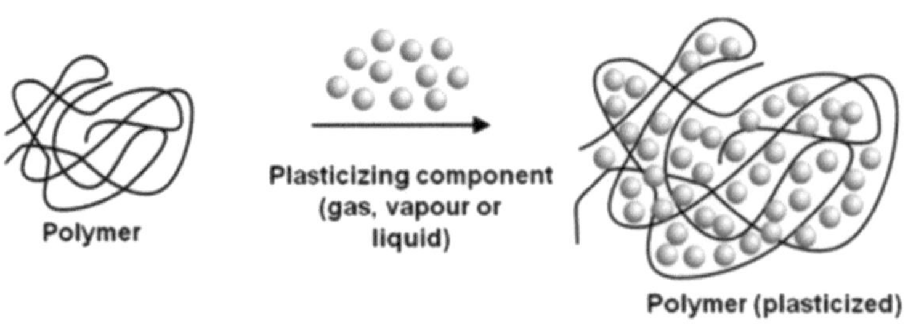

Figure 1.
Plasticization phenomena resulting in facilitated polymer mobility and increased free volume in the polymer, adapted from Kikic et al. [9].

that of plasticization pressure and amount of CO_2 that invoke plasticization, in different glassy polymer classes through measurement of gas permeation and sorption [10]. Kapantaidakis et al. demonstrated accelerated CO_2 plasticization effect in ultrathin polymer structures by measuring gas permeance at increasing operating pressures [13]. Horn and Paul [14] studied the CO_2 plasticization (reversible) and conditioning (non-reversible) effects in thin and thick glassy polymeric membranes to reaffirm conclusion by Kapantaidakis et al. [14]. Tiwari et al. extended the study by Horn and Paul [14] to high free volume glassy perfluoropolymers through evaluation of CO_2 permeability and ellipsometry measurement [15]. Reviews of study related to CO_2 plasticization in different membranes have been provided in works by Suleman et al. [16].

From previous works, it is found that many transport property studies devoted to plasticization include only study of membrane morphology and permeability measurements near ambient conditions. Complementary information on the individual contributions of the sorption and diffusion coefficients to the overall performance at non-ambient and elevated temperatures is rarely reported. Most of the laboratory data have been limited to study of gas transport characteristic within polymeric membrane at ambient operating temperatures (25–35°C). This is because it appears to be not convenient and time consuming to control the operating conditions at different ranges.

Hence, the objective of present study is to study the effect of CO_2 to plasticization of Polysulfone membrane at varying operating conditions. PSF have the most advantages among all polymeric membranes since it easily forms thin film on membrane support surfaces, while demonstrating behaviors such as chemical inertness, good mechanical strength and stable property, which have encouraged their usage in biogas processing. Bos et al. reported a plasticization pressure of 34 bar at 23°C [10] for polysulfone. Nonetheless, the collected data is limited and not extended to elevated pressure, as well as other operating temperatures. In typical biogas processing, the entering gas are in the range of 35–55°C in order to suit the temperature for membrane separation [17]. In our recent work, we studied the interaction of CO_2 with polysulfone membranes at varying operating temperatures and CO_2 concentrations through employment of atomistic simulation technique and concluded that lower operating temperature constituted to more apparent plasticization effect to membrane morphology [18]. Nonetheless, the study has not been extended to study of gas transport property.

Therefore, this work aims to assemble a sequence of experimental procedure to study gas transport property, which includes solubility, diffusivity and permeability, during plasticization at varying operating temperatures. In an overall, firstly PSF membrane has been fabricated through in-house experimental procedure. Then, the PSF membrane has been analyzed through characterization in order to evaluate applicability of developed membrane. Subsequently, the solubility and permeability of noncondensable methane at varying operating temperatures have been measured and validated with published experimental data to determine applicability of the experimental setup. Then, the gas transport property of condensable CO_2 at different operating temperatures has been elucidated to study CO_2 plasticization effect in membrane. Finally, empirical models have been used to quantify gas transport behavior of the gases.

2. Methodology

This section discusses the methodology that has been adapted in current work. The overall workflow is presented in **Figure 2**.

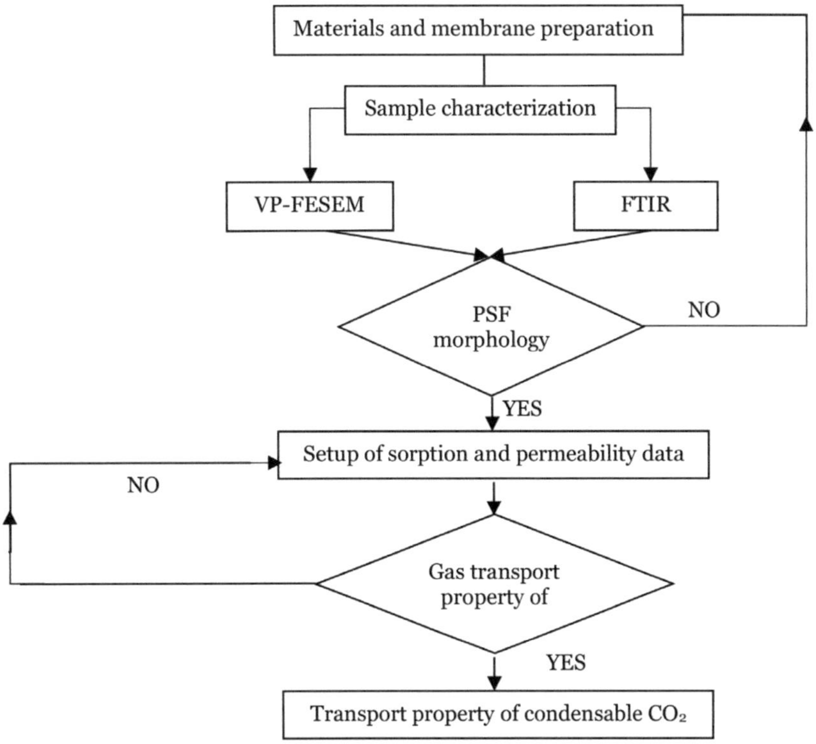

Figure 2.
Process diagram of overall workflow.

This section describes the materials and fabrication methodology to prepare the PSF membrane as well as analysis methodology to validate the developed membrane. In addition, the pressure decay and constant-pressure variable volume methodology for measurement of gas penetrants solubility and permeability have been elaborated.

2.1 Materials and membrane fabrication

The polysulfone (PSF) dense film was prepared via solution casting method using N-Methyl-2-pyyrolidone (NMP) as solvent [19, 20] with a composition of 25 wt% PSF. The PSF was manufactured and supplied in pellet form by Aldrich (MW ~35,000 by light scattering) while NMP from Merck (analytical grade) was used as received. Flow diagram characterizing the chronological procedure for fabrication of PSF dense membrane is depicted in **Figure 3**.

In the beginning, the PSF pellets were dehydrated overnight to get rid of unwanted water content by heating it in a vacuum oven. Subsequently, the amount of dried PSF pellets and filtered NMP solvent was measured prior to mixing them together for 24 hours. When approaching the end of the mixing process, a clear homogenous solution was observed.

Then, an ultrasonication water bath has been employed to desonicate the mixture with a total duration of 4 h before leaving it for 24 h free standing degas. This is aimed to remove any bubbles formed during the mixing protocol while enhancing its homogeneity. The casting solution was then poured into a leveled and clean glass Petri dish, which was covered with aluminum foil to reduce its evaporation rate.

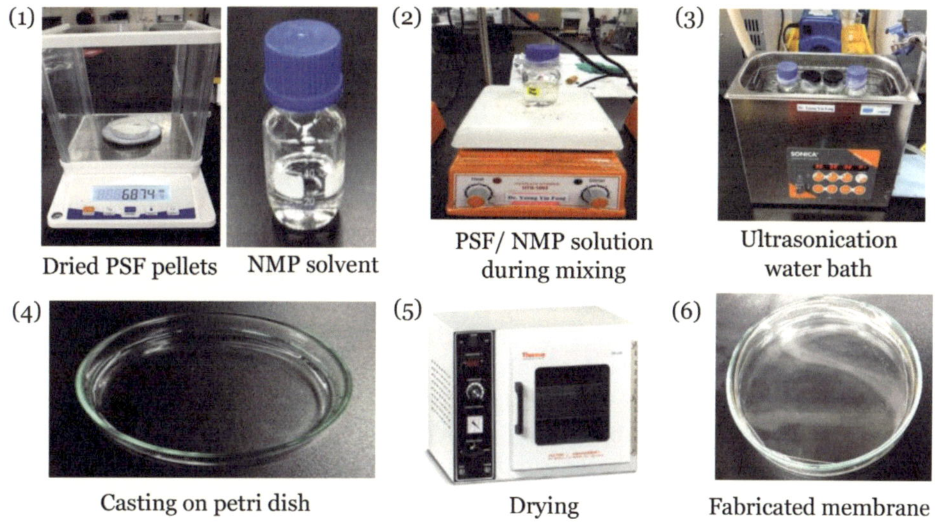

Figure 3.
Flow diagram of procedure for preparation of PSF membrane.

Subsequently, the membranes were subjected to vacuum drying at a heating rate of 20°C/h from 40 to 180°C followed by annealing at 180°C for 24 h. This is to prevent formation of defects in the membrane due to fast evaporation of the solvent. Finally, the PSF membrane film was carefully peeled off from the Petri dish once the cast solution was completely dried.

2.2 Sample characterization

In this section, the characterization analysis that has been used to investigate morphology of the prepared membrane has been discussed. The analysis is crucial to evaluate that the fabricated PSF membrane is dense and defect free. At the same time, it is aimed to ensure that all undesirable solvents that can potentially affect membrane characteristic have been removed accordingly. This is to confirm its applicability prior to measurement of gas solubility and permeability behavior in subsequent sections.

2.2.1 Variable pressure field emission scanning Electron microscope

The variable pressure field emission scanning electron microscope (VP-FESEM, Zeiss Supra 55 VP) was employed to evaluate membrane morphology of the fabricated PSF membrane. Cross sectional side of the membranes were prepared for VP-FESEM analysis via immersion in liquid nitrogen before fracturing the film in order to prevent morphology distortion. All the membrane samples were subsequently sputter coated with platinum using Quorum Q150R S coater prior to imaging. Membrane samples were observed using VP-FESEM with magnification at 500.

2.2.2 Fourier transform infrared spectrometer

Fourier transform infrared spectrometer (FTIR, Perkin Elmer Spectrum One) was operated under transmission with 50 scans in the wavelength range of 450–4000 cm^{-1} to determine IR spectra of the fabricated PSF membrane.

2.3 Solubility measurement (pressure decay methodology)

The principle is based on a dual-chamber pressure decay setup, which has been demonstrated in detailed elsewhere [21]. In this approach, the quantity of gas originally introduced to a sorption system and equilibrated quantity of gas left behind after sorption into a polymer located within the sorption system are determined. This requires measurement in the decline of pressure after sorption of gas into a polymer under study, the temperature of gas, and volume of the system in which the experiment takes place. By measuring the aforementioned variables, the initial and final number of gases existing in the sorption system can be determined directly.

The concentration of gas molecule, x, sorbed within the polymer membrane at any operating temperature has been obtained through Eq. (1), where 22,414 cm^3/mol corresponds to a simple numerical conversion factor and V_p (cm^3) is volume of the polymer sample in the membrane chamber, which has been determined through the conventional fluid displacement method.

$$x = n_p \left(\frac{22414}{V_p} \right) \tag{1}$$

In this study, in order to invoke sorption of polymeric membranes at varying operating temperatures, a constant temperature water bath has been employed. In this context, temperature of the system is consistently controlled at the designated value by submerging the sorption cell within the temperature regulated water bath. Operating temperature is increased gradually from 35 to 55°C with an interval of 10°C for each incremental step. As for pressure increment, it has been continually increased from 5 to 50 bars with an interval of 5 bars to determine sorption isotherm of gases. CH_4 gas has been introduced to the sorption cell first prior to CO_2 since condensable gases can potentially cause irreversible plasticization and swelling effect to the membrane morphology that affects its sorption capability as an end result.

2.4 Permeation measurement

This section describes the experimental setup for gas permeation testing across the PSF membrane. The apparatus adopted a constant-pressure variable volume system to measure gas permeability by measuring the permeate flow rate at atmospheric downstream pressure using a bubble flow meter. Schematic diagram for gas permeation measurement has been provided in **Figure 4**.

The system consists of a feed inlet point, a pressure regulation system and a mass flow controller. The amount of gas from feed inlet point was controlled using a mass flow controller. The permeation apparatus is developed for high pressure testing devoted to CO_2 and CH_4. For high pressure applications, all fittings and valves were supplied by Swagelok with pressure-rating > 70 bar while all sensors are capable to read a maximum pressure of 100 bars. In addition, operating temperature of the membrane has been controlled at constant and designated value by regulating oven temperature. Before conducting experiment, the system has been evacuated with a vacuum pump overnight to eliminate any gas or vapors in the system. Leak tests have been performed after degassing process to ensure that the equipment is safe before experiment proceeded.

For polymer structure, it is important to heat it ~10°C above its glass transition temperature, T_g, in the absence of any mechanical stress to erase all previous thermal history as well as to relax any molecular orientation captured during film formation [22]. The thermal history removal protocol has been adapted from

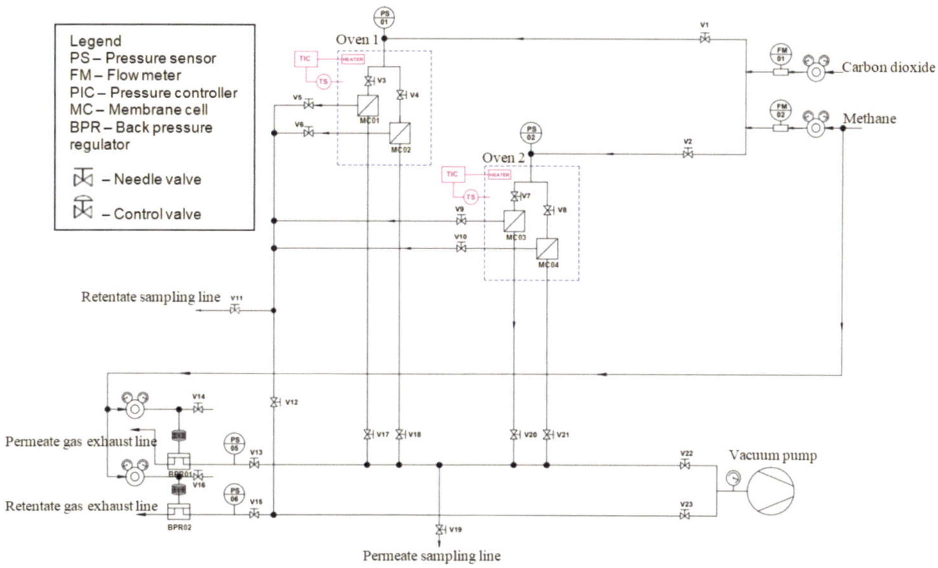

Figure 4.
Parallel membrane cell for reproducibility of gas permeation.

Huang et al. work [22]. The procedure has been conducted in the vacuum oven for 30 min to prevent any oxidation since equilibration at the rubbery state should be tentatively achieved over this time span based on Struik's report [23]. After heating, the polymer membrane has been immediately removed from the vacuum oven and has been quenched to ambient temperature, while preparing the membrane for permeation test.

For the separation process, the membrane sample of 3.14 cm^2 effective area has been mounted in a membrane test cell to study the separation efficiency at various operating conditions. The membrane area has been constituted by cautiously locating the membrane films on aluminum tape over a circle hole with a diameter of 1 cm while avoiding any folding that destroys the membrane surface. Finally, a second piece of aluminum tape and Whatman® Anodisc filter has been adhered to the underside of the membrane for mechanical support to withstand a wide range of operating pressure.

The process of mounting the membrane within the test cell has been conducted within 15 min before bringing the polymer membrane to the desired operating temperature, which are 35, 45 and 55°C respectively. The pressure has been increased gradually from atmospheric condition to a maximum of 50 bars with an increment step of 5 bars. Upstream gas at required operating pressure, temperature and flow rate has been introduced into the membrane for permeation test. Volumetric permeation rates in the permeate stream has been determined with a soap bubble flow meter. Lastly, the entire system should be evacuated to fully degas the system before proceeding to other experiment.

At least three measurements were performed to evaluate the flowrate and composition during membrane separation process. Since the current study addresses high pressure conditions, non-ideal gas conditions should be considered. The driving force for this case is described as the distinction in fugacity from the high to low end across the membrane. A nonideal equation of state has been employed to compute fugacity of CO_2 and CH_4 on the feed side. On the other hand, since the permeate pressure is remained at atmospheric condition, the nonideality associated to real gas behavior can be disregarded. The permeability of gas component i (barrer), P_i, is calculated based on Eq. (2).

$$P_i = \frac{tV_P}{A_m(f_h - f_l)} \quad (2)$$

In Eq. (2), V_P is the permeate flow rate (cm^3 (STP)/s), t is the thickness of membrane (cm), A_m is the membrane area (cm^2), f_h and f_l are the fugacities in feed side and permeate side respectively (cmHg), subscript i denotes CO_2 or CH_4. The permeability of the membrane is expressed in the unit of Barrer (1 Barrer = 1×10^{-10} cm^3 (STP) cm/s cm^2 cmHg).

3. Results and discussion

In this part, the results pertaining to experimental section from the fabricated PSF polymeric membrane, gas transport properties of incondensable CH_4 and condensable CO_2 within the PSF membrane and empirical model to quantify the permeation behavior have been discussed.

3.1 Membrane morphology

To understand the membrane morphology, several characterization methodologies have been conducted to analyze the fabricated membrane, which comprised those of VP-FESEM and FITR, as discussed in the subsections.

3.1.1 VP-FESEM

The structure of dense PSF membrane at a magnification of 500 is depicted in **Figure 5**.

The PSF membrane is consisted of a dense, nonporous and single polymer layer that is homogenous in all directions. The thickness of the membrane is ~78 μm. The smooth membrane configuration without defects ensures its applicability for solubility and gas permeation measurement in subsequent section.

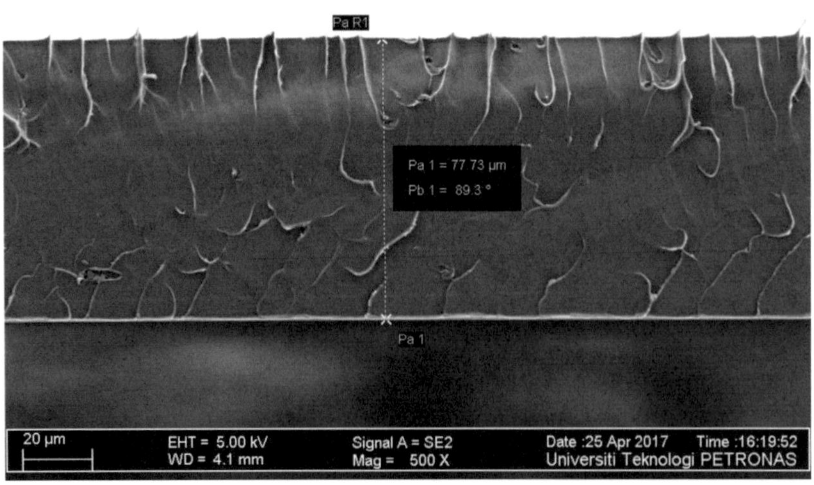

Figure 5.
Cross sectional of PSF dense membrane.

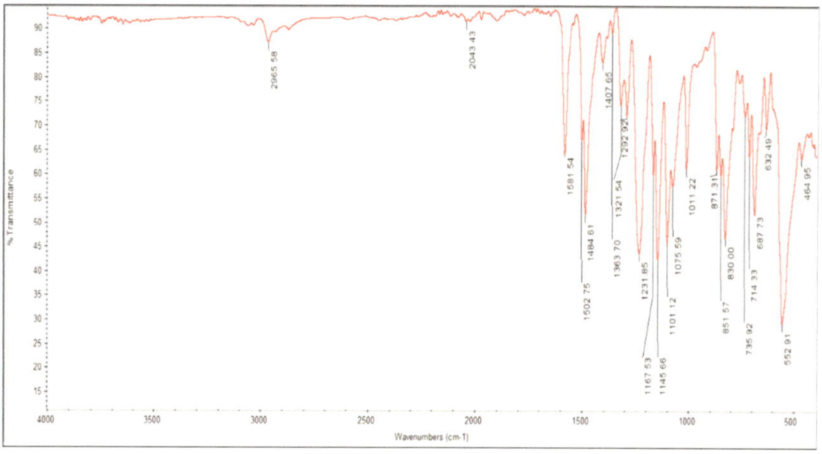

Figure 6.
FTIR spectrum of synthesized PSF membrane.

3.1.2 FTIR

FTIR is the most effective alternative to elucidate the functional group of membrane. **Figure 6** depicts the IR spectrum of the PSF polymeric membrane obtained in present work.

For PSF membrane, the peak at 2965.58 cm^{-1} correspond to stretching vibration of asymmetric and symmetric C-H bond. On the other hand, the peaks at 1581.54 and 1484.61 cm^{-1} represent the C=C bond in the PSF repeat chain. The IR spectra peak observed at 1407.65 and 1363.70 cm^{-1} correspond to the asymmetric and symmetric C-H bending deformation of methyl group. Amine stretching is depicted at 1101.12 cm^{-1}, while phenyl ring substitution band is noticed at 851.57, 830.00, 714.33 and 687.73 cm^{-1}. IR spectrum noticeable at 1231.85 cm^{-1} represents the presence of asymmetric C-O-C stretching by aryl ether group. The peaks at 1167.53 and 1101.12 cm^{-1} are assigned to asymmetric and symmetric O=S=O stretching of sulfonate group. In addition, the peak at 1407.65 cm^{-1} has been attributed to stretching vibration of aromatics in PSF. All the functional groups are consistent to those observed in the repeat unit of Polysulfone [18]. The good accordance demonstrates the validity of the synthesized PSF membrane and elimination of any impurities/solvent that can potentially affect the membrane separation performance.

3.2 Validation with transport properties of methane

To validate applicability of the sorption laboratory setup, gas methane has been introduced to the PSF membrane with operating pressure at an incremental step, to acquire the sorption behavior of CH_4 in PSF membrane, as shown in **Figure 7**. Experimental data by Sada et al. that studied the effect of operating temperature to the solubility of CH_4 within PSF has also been provided as Ref. [24]. The sorption experimental data obtained from current work exhibits close agreement with published results by Sada et al. [24]. The good compliance suggests that the fabricated PSF polymeric membrane and experimental setup are of adequate soundness to produce defects and error free experimental results.

From **Figure 7**, it is shown that the concentration of CH_4 being sorbed into the polymeric membrane is enhanced when operating pressure increases at all operating temperature. The increment in CH_4 concentration can be explained through greater driving force that advances the sorption of gas molecules within the free

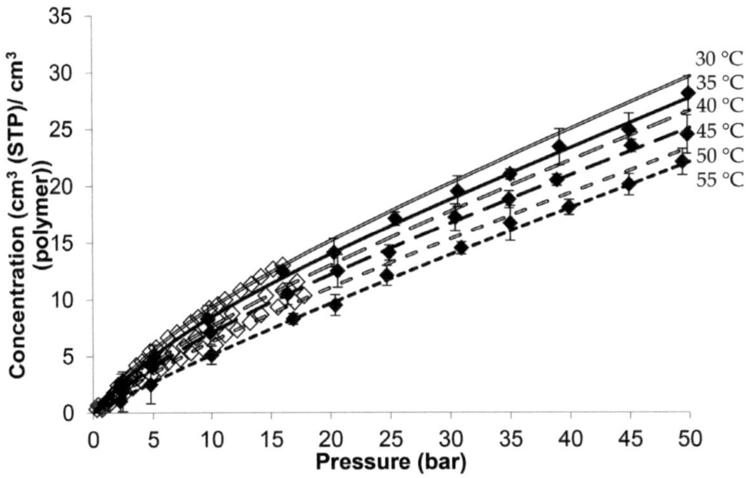

Figure 7.
*CH_4 sorption isotherm for polysulfone. [♦ In-house collected sorption data ◇ Sorption data by Sada et al. [24]. Close line - Prediction of dual mode sorption model by Eq. (3) with parameters in **Table 1** for in-house collected sorption data Open line - Prediction of dual mode sorption model by Eq. (3) with parameters in **Table 1** for sorption data by Sada et al. [24]. (——— 30/35°C, – – – - 40/45°C, - - - - - · 50/55°C)].*

spaces of polymeric membrane matrix. The sorption of CH_4 gas molecules is found to be decreasing when operating temperature is increased [18]. The decrement is intuitively reasonable since gas molecules have higher affinity to remain in the gaseous state than rather being sorbed into the membrane at higher operating temperature. In addition, the sorption isotherm of CH_4 is found to exhibit good correlation to the dual mode sorption model as depicted in Eq. (3).

$$C_i = C_{Di} + C_{Hi} = k_{Di}f_i + \frac{C'_{Hi}b_i f_i}{1 + b_i f_i} \tag{3}$$

The dual mode sorption model suggests that the total concentration of gas *i* in a polymer matrix is composed of two idealized molecular scale environment, in which C_i is the total concentration of gas in the polymer; C_{Di} is equilibrium population existing in the polymer matrix under the dissolved mode and is governed by Henry's Law equation, while C_{Hi} is the non-equilibrium population existing in excess within the hole-filling environment governed by Langmuir parameters [25–27]. Moreover, k_{Di} is the Henry's law coefficient that characterizes dissolution of a pure gas, *i*, in the polymer, b_i and C'_{Hi} is the Langmuir hole affinity parameter and the capacity parameter respectively, while f_i is fugacity of the gas system [26, 28]. The fitted dual mode sorption parameters are provided in **Table 1**, which has been summarized alongside the reported values by Sada et al. [24].

In has been demonstrated from **Table 1** that the parameters are in satisfactory agreement with one another, attributed to the small distinction of the solubility

Temperature (°C)	k_{Di} (cm³(STP) cm$^{3-1}$ bar¹)	b_i (bar^{-1})	C'_{Hi} (cm³ (STP) cm$^{3-1}$)
35	0.4123 (0.4352 @ 30°C)[a]	0.1055 (0.1145 @ 30°C)[a]	8.45 (9.26 @ 30°C)[a]
45	0.3859 (0.4076 @ 40°C)[a]	0.0812 (0.0874 @ 40°C)[a]	7.21 (7.63 @ 40°C)[a]
55	0.3589 (0.3711 @ 50°C)[a]	0.0678 (0.0738 @ 50°C)[a]	5.34 (6.03 @ 50°C)[a]

[a]The number in bracket is the experimental value by Sada et al. [24].

Table 1.
Dual-mode sorption parameters for methane in polysulfone film as a function of operating temperature.

characteristics as a whole. The reported values are found to be consistently higher for lower operating temperature attained through higher sorption capacity as explained earlier. The good compliance with previous published literatures and fit to the commonly employed dual mode sorption model demonstrates that the dual mode sorption cell is of high accuracy for plasticization study in subsequent section.

Similarly, validity of the gas permeation cell has been investigated by comparing measured methane permeability data with published experimental results by Sada et al. [24], such as that shown in **Figure 8** whereby a close agreement has been obtained in between the two. Measured permeabilities for methane in polysulfone films are illustrated as a function of upstream gas pressure in **Figure 8**. At every temperature, the mean permeability coefficients were found to decrease with an increase in upstream pressure. Such pressure dependence seems to be characteristic of glassy polymers. The gas permeability is found to be consistently higher at greater operating temperature. The contributing factor is free volume within the structure of the polymer has increased as the temperature is further increased, while gaining additional energy to execute diffusional jump.

The permeability of gas through a glassy polymeric membrane is frequently characterized through the partial immobilization model [29], which has evolved from the dual mode sorption model in Eq. (3), such as that presented in Eq. (4).

$$P_i = k_{D,i} D_{0,i} \left[1 + \frac{F_i C'_{Hi} b_i}{1 + b_i f_i} \right] \tag{4}$$

In Eq. (4), $k_{D,i}$, C'_{Hi} and b_i are parameters from dual mode sorption model, while $D_{0,i}$ and F_i represent the diffusion coefficient in the limit when concentration of the mobile gas $C_{M,i} \rightarrow 0$ and the ratio of the diffusivity through the microvoids to that through the polymeric matrix. The additional parameters in the model have been summarized in **Table 2**.

For these results, $D_{0,i}$ and F_i appear to be a function of temperature. It is found that $D_{0,i}$ increases with increment in temperature, which has been rationalized through the enhancement in diffusion energy. A small F_i value corresponds to a relatively low diffusivity through the Langmuir regions. At lower operating

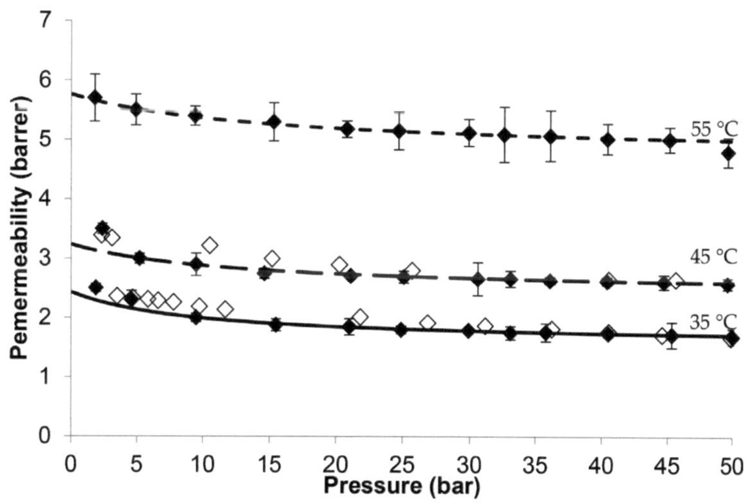

Figure 8.
*CH_4 gas permeability data for polysulfone. [♦ In-house permeability data ◇ Permeability data by Sada et al. [24]. Line - Prediction of partial immobilization model by Eq. (4) with parameters in **Table 2** for in-house collected permeability (———35°C, – – – -45°C, - - - - - · 55°C)].*

Temperature (°C)	$D_{0,i}$ (cm^2/s)	F_i
35	2.87×10^{-8}	0.607
45	4.75×10^{-8}	0.554
55	9.98×10^{-8}	0.507

Table 2.
Partial immobilization parameters for methane in polysulfone film, model adapted from Scholes et al. [29].

temperature, the Langmuir microvoids exist in a large number with greater sizes [29]. Therefore, gas molecules have a higher tendency to be transported through the Langmuir regions with lower resistance. When operating temperature is further increased, there is a reduction in the number and size of Langmuir microvoids, which consequently restraints the transport in such region. Therefore, there is a shift from dominancy of Langmuir to Henry's region with increment in operating temperature, which contributes to a smaller F_i value. In a similar manner, the satisfactory compliance with Sada et al. published literatures and fit to the commonly employed partial immobilization model demonstrates that the permeation cell rig is of high accuracy for measurement of CO_2 plasticization study in next section.

3.3. CO_2 plasticization in PSF membranes

CO_2 sorption in PSF membrane has been measured with increment in operating pressure at varying operating temperatures, such as that provided in **Figure 9**.

Published literature data by Sada et al. for solubility of CO_2 within PSF has also been provided [24]. In an overall, it is depicted that the collected sorption data of present study is not substantially different from the reported values by Sada et al. [24]. The sorption data of present work is consistently higher than that reported by Sada et al. at different operating temperatures, which can be deduced via the difference in source of polysulfone to prepare the membrane samples. The sorption data also demonstrates a good agreement with the dual mode sorption model, with close compliance with that reported by Sada et al. [24], such as that summarized in **Table 3**.

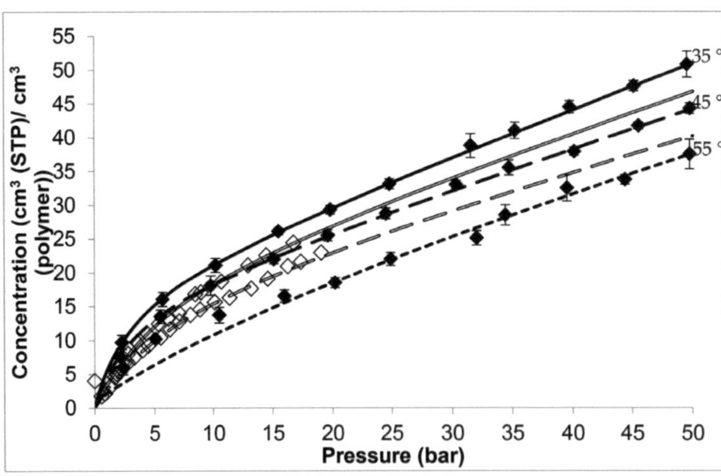

Figure 9.
*CO_2 sorption isotherm for polysulfone. [♦ In-house collected sorption data ◇ Sorption data by Sada et al. [24]. Close line - Prediction of dual mode sorption model by Eq. (3) with parameters in **Table 3** for in-house collected sorption data Open line - Prediction of dual mode sorption model by Eq. (3) with parameters in **Table 3** for sorption data by Sada et al. [24] (——— 35°C, – – – - 45°C, - - - - - · 55°C)].*

Temperature (°C)	k_{Di} (cm^3(STP) cm$^{3-1}$ bar^1)	b_i (bar^{-1})	C'_{Hi} (cm^3 (STP) cm$^{3-1}$)
35	0.6748 (0.5872)[b]	0.3678 (0.1757)[b]	18.20 (19.40)[b]
45	0.5840 (0.5014)[b]	0.3415 (0.1530)[b]	15.93 (17.1)[b]
55	0.4932	0.3152	13.67

[b]*The number in bracket is the experimental value by Sada et al. [24].*

Table 3.
Dual-mode sorption parameters for carbon dioxide in polysulfone film as a function of operating temperature.

As for gas permeability of CO_2 through PSF membrane at varying pressures, it has been tabulated in **Figure 10**. Similarly, applicability of the data has been demonstrated through good compliance with published experimental data by Sada et al. [24].

It is found that gas permeability experiences a decrement before reaching the plasticization pressure at 34.9, 36.1 and 38.0 bars respectively for operating temperature of 35, 45 and 55°C. This has been attributed to rapid decrement in gas solubility when the sorption level off at high pressure due to saturation of favorable sites. Nevertheless, beyond the plasticization pressure, an increase in permeability has been observed because the diffusion coefficient increases with pressure much more rapidly than the solubility coefficient that decreases with pressure, which has been elucidated through plasticization effect that enhances the diffusivity of gas molecules to a large extend when the polymeric membrane is swelled. Viewing from the impact of plasticization pressure, it is shifted to higher value at greater operating temperature. This has been attributed to lower sorption of condensable CO_2 when the gas has the tendency to maintain at its gas state with increment in temperature.

The parameters for partial immobilization model of CO_2 have been summarized in **Table 4** with a similar trend observed to that for methane. Nonetheless, the parameters are only applicable to condition before the plasticization pressure is met. After that, the plasticization behavior has been characterized through Eq. (5) that describes permeability of gas within the glassy membrane undergoing plasticization [29].

$$P_i = \frac{D_{0,i}}{\beta_i f_i}\left[\exp\left(\beta_i k_{Di} f_i\left(1 + \frac{F_i C'_{Hi} b_i}{k_{Di}(1 + b_i f_i)}\right)\right) - 1\right] \quad (5)$$

Variables describing the modified partial immobilization model for plasticized membrane are provided in **Table 4** as well.

Regardless of demonstrating a similar trend of increment in value with increasing temperature, the $D_{0,i}$ values after plasticization are found to be relatively smaller as compared to its counterpart with pristine unaltered PSF structure. This has been attributed to a smaller amount of mobile gas when the favorable sites become concentrated and occupied. With respect to temperature dependency, β_i is found to decrease with increment in temperature, which implies that the plasticization potential reduces with temperature. It has been proposed that polymeric membrane experiences a decrement in Langmuir microvoids with increment in operating temperature [29]. Therefore, there are fewer pathways for CO_2 to interact with functional group of the polymeric chains, which consequently reduces the plasticization potential at higher temperature. In addition, the condensable CO_2 also has a higher tendency to exist in the gaseous state, which reduces its plasticization power when operating temperature is increased.

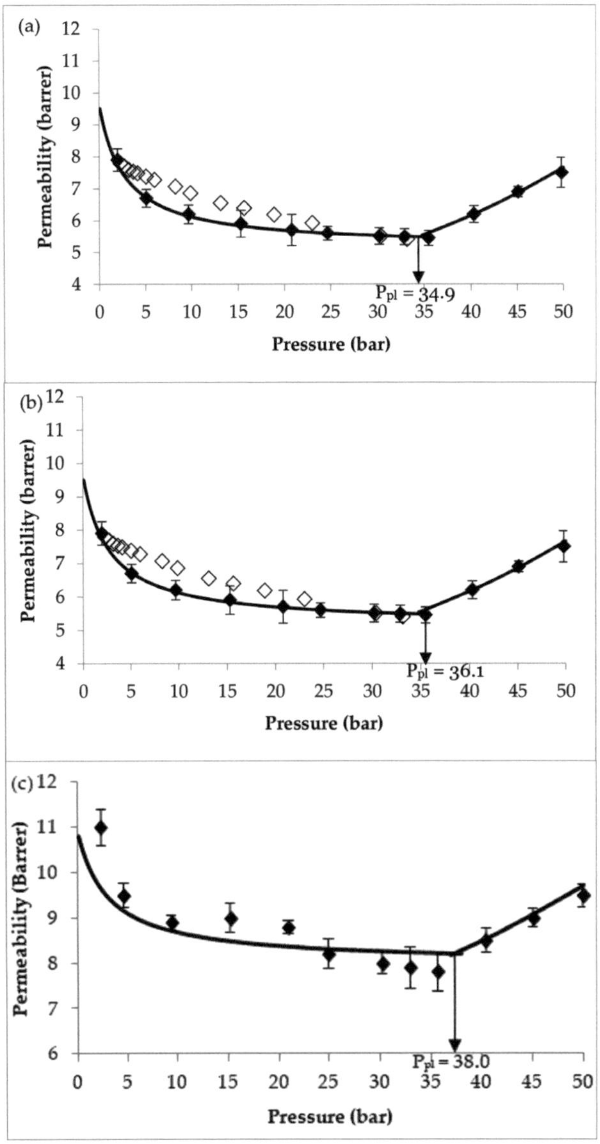

Figure 10.
*CO_2 gas permeability for polysulfone at (a) 35°C, (b) 45°C and (c) 55°C. [♦ In-house permeability data ◇ Permeability data by Sada et al. [24]. Line - Prediction of partial immobilization model by Eq. (4) and Eq. (5) with parameters in **Table 4** for in-house collected permeability].*

Temperature (°C)	$D_{0,i}$ ($\times 10^{-8}$ cm^2/s)		F_i	β_i
	Before plasticization	After plasticization		
35	5.73	2.61	0.1307	0.0537
45	8.33	4.31	0.1107	0.0506
55	12.1	7.02	0.0856	0.0491

Table 4.
Partial immobilization parameters for CO_2 in polysulfone film as a function of operating temperature, model adapted from Scholes et al. [29].

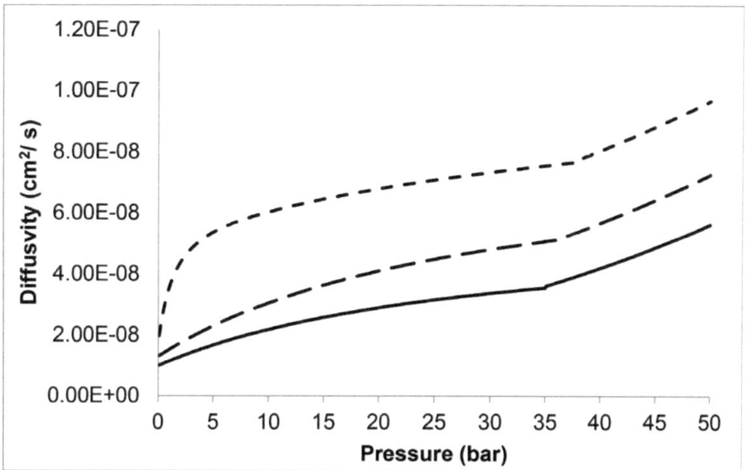

Figure 11.
CO_2 diffusivity for polysulfone (——— 35°C, – – – - 45°C, - - - - - · 55°C).

As for gas diffusivity shown in **Figure 11**, prior to CO_2 plasticization effect, the value at a low pressure is relatively lower, because most of the gas molecules are in the Langmuir mode and it has been reported that gas molecules sorbed into the Henry's mode sites inherit greater diffusivity than its counterpart [26].

The apparent diffusivity increases and reaches the asymptotic limit of diffusivity of Henry's Law proportion at high pressure. Nonetheless, it is found that after the plasticization pressure, gas diffusivity increases exponentially when pressure is further increased. The observation can be explained through enhanced interaction between CO_2 gas molecule and polymeric matrix, which contributes to augmented swelling and increment in free volume that forms pathway for diffusion of gas.

4. Conclusions

In present study, in house experimental work and setup has been conducted to fabricate, to characterize and to evaluate the gas transport properties in polysulfone (PSF) membrane film, typically those with plasticization characteristic. Validity of the solubility and gas permeability measurement has been demonstrated through good accordance with published experimental results and satisfactory empirical fitting to the dual mode sorption and partial immobilization models, which are well-known equations to quantify gas sorption and permeation in glassy polymeric membranes. To conclude, polysulfone membranes have permeability–pressure and concentration-pressure isotherms that vary with temperature. The plasticization potential decreases with temperature, implying that CO_2 ability to plasticize the polysulfone membrane reduces at higher temperature. In addition, the plasticization pressure is shifted to higher value with increment in temperature (34.9 bars at 35°C to 36.1 bars at 45°C to 38.0 bars at 55°C). In addition, gas permeability is found to be enhanced at greater operating temperature, which can be rationalized through greater activation energy to execute diffusional jump in increased free volume structure. From findings of present study, it is found that higher operating temperature is favorable for membrane operation since it promotes gas permeation, which enables more efficient removal of CO_2 from biogas under the same membrane area requirement. In addition, higher operating temperature also suppresses the effect of plasticization by exhibiting higher plasticization pressure. The study of CO_2

plasticization at varying operating temperatures is anticipated to be extended to mixed CO_2/CH_4 system to verify the behavior in real membrane gas separation.

Acknowledgements

This work is done with the financial support from Universiti Teknologi PETRONAS.

Conflict of interest

The authors declare that there is no conflict of interest.

Author details

Serene Sow Mun Lock[1], Kok Keong Lau[1]*, Azmi Mohd Shariff[1], Yin Fong Yeong[1] and Norwahyu Jusoh[2]

1 CO_2 Research Center (CO_2RES), Department of Chemical Engineering, Universiti Teknologi PETRONAS, Seri Iskandar, Malaysia

2 Centre for Contaminant Control (CenCo), Department of Chemical Engineering, Universiti Teknologi PETRONAS, Seri Iskandar, Malaysia

*Address all correspondence to: laukokkeong@utp.edu.my

References

[1] Song Z, Zhang C, Yang G, Feng Y, Ren G, Han X. Comparison of biogas development from households and medium and large-scale biogas plants in rural China. Renewable and Sustainable Energy Reviews. 2014;**33**:204-213. DOI: 10.1016/j.rser.2014.01.084

[2] Yépez-García RA, Dana J. Mitigating Vulnerability to High and Volatile Oil Prices: Power Sector Experience in Latin America and the Caribbean. Washington: The World Bank; 2012. DOI: 10.1596/978-0-8213-9577-6

[3] Hart A, Gnanendran N. Cryogenic CO_2 capture in natural gas. Energy Procedia. 2009;**1**:697-706. DOI: 10.1016/j.egypro.2009.01.092

[4] Hao J, Rice PA, Stern SA. Upgrading low-quality natural gas with H_2S- and CO_2-selective polymer membranes: Part I. Process design and economics of membrane stages without recycle streams. Journal of Membrane Science. 2002;**209**:177-206. DOI: 10.1016/S0376-7388(02)00318-6

[5] Gori G, Gabetta G. The use of knowledge management to improve pipeline safety. In: Bolzon G, Boukharouba T, Gabetta G, Elboujdaini M, Mellas M, editors. Integrity of Pipelines Transporting Hydrocarbons: Corrosion, Mechanisms, Control, and Management. Netherlands: Springer; 2010. pp. 1-16. DOI: 978-94-007-0588-3

[6] Marriott J, Sørensen E. A general approach to modelling membrane modules. Chemical Engineering Science. 2003;**58**:4975-4990. DOI: 10.1016/j.ces.2003.07.005

[7] Jung HJ, Han SH, Lee YM, Yeo Y-K. Modeling and simulation of hollow fiber CO_2 separation modules. Korean Journal of Chemical Engineering. 2011;**28**:1497-1504. DOI: 10.007/s11814-010-0530-y

[8] Coleman MR, Koros WJ. Conditioning of fluorine containing polyimides. 1. Effect of exposure to high pressure carbon dioxide on permeability. Macromolecules. 1997;**30**:6899-6905. DOI: 10.1021/ma961323b

[9] Kikic I, Vecchione F, Alessi P, Cortesi A, Eva F, Elvassore N. Polymer plasticization using supercritical carbon dioxide: Experiment and modeling. Industrial & Engineering Chemistry Research. 2003;**42**:3022-3029. DOI: 10.1021/ie020961h

[10] Bos A, Pünt IGM, Wessling M, Strathmann H. CO_2-induced plasticization phenomena in glassy polymers. Journal of Membrane Science. 1999;**155**:67-78. DOI: 10.1016/S0376-7388(98)00299-3

[11] Wessling M, Schoeman S, van der Boomgaard T, Smolders CA. Plasticization of gas separation membranes. Gas Separation & Purification. 1991;**5**:222-228. DOI: 10.1016/0950-4214(91)80028-4

[12] Houde AY, Kulkarni SS, Kulkarni MG. Permeation and plasticization behavior of glassy polymers: A WAXD interpretation. Journal of Membrane Science. 1992;**71**:117-128. DOI: 10.1016/0376-7388(92)85011-7

[13] Kapantaidakis GC, Koops GH, Wessling M, Kaldis SP, Sakellaropoulos GP. CO_2 plasticization of polyethersulfone/polyimide gas-separation membranes. AIChE Journal. 2003;**49**:1702-1711. DOI: 10.002/aic.690490710

[14] Horn NR, Paul DR. Carbon dioxide plasticization and conditioning effects in thick vs. thin glassy polymer films. Polymer. 2011;**52**:1619-1627. DOI: 10.016/j.polymer.2011.02.007

[15] Tiwari RR, Smith ZP, Lin H, Freeman BD, Paul DR. Gas permeation in thin films of "high free-volume" glassy perfluoropolymers: Part II. CO_2 plasticization and sorption. Polymer. 2015;**61**:1-14. DOI: 0.1016/j.polymer.2014.12.008

[16] Suleman MS, Lau KK, Yeong YF. Plasticization and swelling in polymeric membranes in CO_2 removal from natural gas. Chemical Engineering & Technology. 2016;**39**:1604-1616. DOI: 10.002/ceat.201500495

[17] Safari M, Ghanizadeh A, Montazer-Rahmati MM. Optimization of membrane-based CO_2-removal from natural gas using simple models considering both pressure and temperature effects. International Journal of Greenhouse Gas Control. 2009;**3**:3-10. DOI: 1016/j.ijggc.2008.05.001

[18] Lock SSM, Lau KK, Shariff AM, Yeong YF, Bustam MA, Jusoh N, et al. An atomistic simulation towards elucidation of operating temperature effect in CO_2 swelling of polysulfone polymeric membranes. Journal of Natural Gas Science and Engineering. 2018;**57**:135-154. DOI: 10.1016/j.jngse.2018.07.002

[19] Wijenayake SN, Panapitiya NP, Versteeg SH, Nguyen CN, Goel S, Balkus KJ, et al. Surface cross-linking of ZIF-8/polyimide mixed matrix membranes (MMMs) for gas separation. Industrial & Engineering Chemistry. 2013;**52**:6991-7001. DOI: 10.1021/ie400149e

[20] Jusoh N, Yeong YF, Lau KK, Shariff MA. Enhanced gas separation performance using mixed matrix membranes containing zeolite T and 6FDA-durene polyimide. Journal of Membrane Science. 2017;**525**:175-186. DOI: 10.1016/j.memsci.2016.10.044

[21] Stern SA, De Meringo AH. Solubility of carbon dioxide in cellulose acetate at elevated pressures. Journal of Polymer Science Part B: Polymer Physics. 1978;**16**:735-751. DOI: 10.1002/pol.978.180160415

[22] Huang Y, Paul DR. Effect of film thickness on the gas-permeation characteristics of glassy polymer membranes. Industrial & Engineering Chemistry. 2007;**46**:2342-2347. DOI: 10.1021/ie0610804

[23] Struik LCE. Physical aging in amorphous glassy polymers. Annals of the New York Academy of Sciences. 1976;**279**:78-85. DOI: 10.1111/j.749-6632.1976.tb39695.x

[24] Sada E, Kumazawa H, Xu P, Nishigaki M. Mechanism of gas permeation through glassy polymer films. Journal of Membrane Science. 1988;**37**:165-179. DOI: 10.1016/S0376-7388(00)83070-7

[25] Barrer RM, Barrie JA, Slater J. Sorption and diffusion in ethyl cellulose. Part III. Comparison between ethyl cellulose and rubber. Journal of Polymer Science. 1958;**27**:177-197. DOI: 10.1002/pol.958.1202711515

[26] Koros WJ, Paul DR, Rocha AA. Carbon dioxide sorption and transport in polycarbonate. Journal of Polymer Science. 1976;**14**:687-702. DOI: 10.1002/pol.976.180140410

[27] Lock SSM, Lau KK, Shariff AM, Yeong YF, Bustam MA. Thickness dependent penetrant gas transport properties and separation performance within ultrathin polysulfone membrane: Insights from atomistic molecular simulation. Journal of Polymer Science Part B: Polymer Physics. 2018;**56**:131-158. DOI: 10.1002/polb.24523

[28] Story BJ, Koros WJ. Comparison of three models for permeation of

CO_2/CH_4 mixtures in poly(phenylene oxide). Journal of Polymer Science. 1989;**27**:1927-1948. DOI: 10.002/polb.89.090270910

[29] Scholes CA, Chen GQ, Stevens GW, Kentish SE. Plasticization of ultra-thin polysulfone membranes by carbon dioxide. Journal of Membrane Science. 2010;**346**:208-214. DOI: 10.1016/j.memsci.2009.09.036

Section 6

Case Study

Chapter 11

Biofuel Development in Sub-Saharan Africa

Olatunde Samuel Dahunsi, Ayoola Shoyombo and Omololu Fagbiele

Abstract

The quest for renewable and sustainable energy generation is fast becoming widespread across Africa due to the understanding that there is a need to seek an alternative to fuels of fossil origin, which currently sustains the largest portion of the world's energy need. Research into the generation of renewable fuels had been on-going in continents like Europe, South America, Asia, and other developed countries bearing in mind the extinction nature of fossil fuels. Globally, attentions are being drawn to fuel generation from biomass and its derivatives such as lignin, triglycerides, cellulose, and hemicelluloses. The aim is to use such fuels for cooking and heating and in vehicles, jet engines, and other applications. Therefore, the integration of the African continent in the race for biofuel production is germane in the quest for survival and developments considering favorable factors like climate, soil, and land mass among other environmental-friendly resources in different African countries.

Keywords: Africa, biogas, biomass, environment, microorganism

1. Introduction

Environmental pollution by solid wastes and lack of access to adequate energy resources are some of the major challenges facing the human populace in Sub Saharan Africa [1–14]. Out of 21 Sub-Saharan African countries, less than 10% have access to energy [15]. Therefore, there is serious need to search for alternative and renewable energy sources from locally available resources in the quest for human survival and national development in the region [15–18]. Besides, there is a need for the adoption of appropriate and economically feasible technologies for the effective management of solid and liquid wastes and energy recovery from them [19, 20].

The global quest for environmentally friendly and ecologically balanced and sustainable energy has been on the increase over the last few decades and this has forced the world to search for other alternate sources of energy [21, 22]. Besides, one of the major tools for national and international development is energy. Developing countries such as Nigeria depend heavily on fuels from fossil origin. There are enormous conventional energy resources (crude oil, tar sands, natural gas and coal) in Sub-Saharan Africa besides the huge amount of renewable/sustainable energy resources including hydro, solar, wind, biomass, etc.

However, the alternative energy sources demand immense economic investment and technical power to operate, and this makes it little difficult for these countries. Presently, energy from biogas is a reliable, abundant, accessible and economically feasible source of alternative and renewable energy which can be generated using agricultural, domestic and industrial materials employing simple technology [23]. The prospect of this technology is bright because it can be utilized to provide energy for households, rural communities, farms, and industries [18].

Biomass such as perennial grasses has been extensively utilized for biofuel production the world over paramount among which are *Panicum virgatum*, *Miscanthus* species, *Phalaris arundinacea* and *Arundo donax* [24]. The use of *Miscanthus* as an energy grass has attracted attention among the perennial C4 grasses since it has been identified as a perfect energy grass and produces maximally when harvested dry. Yields of 3–10 years old plantations grown in two countries in Europe are 113–30 t/ha. This means that if a yield of 20 t/ha could be achieved; it would produce a total energy yield that is equal to 7 t/ha of oil over the life of each harvest. Switch grass has an energy value that is similar to wood yet with minimal water content [25]. After proper investigation of some crops which were perennial grasses, switch grass was observed to produce the highest potential. Other than staying away from the competition between food and fuel crop usage, they are considered to have energy, financial, and ecological advantages over food crops for certain bioenergy products [25]. These grasses possess qualities and prospects as for their utilization and enhancement as lignocellulosic feedstock. In order to meet up to the large demand of biomass supply, an extensive environmental capacity is to be considered which marginal soils are included [26]. Another nutrient rich grass is Napier grass (*Pennisetum purpureum*), a grass that grows in the tropics and can withstand dry conditions. It has 30.9% total carbohydrates, 27% protein, 14.8% lipid 14.8%, and 9.1% fiber (dry weight). Thus, it is cultivated for livestock as energy crops and it is easy to cultivate with a high productivity rate of 87 ton/ha/year [24]. The feasibility of biogas production from Napier grass was observed and was reported that the methane content, yield and production rate were 53%, 122.4 mL CH_4/g TVS remove, 4.8 mL/h at the optimum condition [26].

2. Rationale for biofuel production in Sub-Saharan Africa

The quest for renewable and sustainable energy generation is fast becoming widespread across Sub-Saharan Africa due to the understanding that there is a need to seek an alternative to fuels of fossil origin which currently sustains the world's-energy need. Research into the generation of renewable fuels had been on-going in continents like Europe, South America, Asia and other developed countries bearing in mind the extinction nature of fossil fuels. Globally, attentions are been drawn to fuel generation from biomass and their derivatives such as lignin, triglycerides, cellulose, and hemicelluloses. The aim is to use such fuels for cooking, heating, as fuels in vehicles, jet engines, and other applications. Therefore, the integration of the African continent in the race for biofuel production is germane in the quest for survival and developments considering present and favorable factors like climate, soil, land mass among other environmental-friendly resources in different Sub-Saharan African countries [28]. Africa is the second largest continent in the world after Asia making up 10% of the world's population which is equivalent to about 80% of the population in India sub-continent [29]. As such, biofuels especially biogas, biodiesel, and bioethanol are being considered as the most potent alternatives to fossil fuels in the continental energy mix [30, 31].

3. Various biofuels produced from lignocelluloses

3.1 Biogas

There are two broad processes in biogas development and these are first, the actual production from both edible and non-edible sources and secondly, the compatible technologies for the fuel usage. Nowadays, large scale biofuel projects are gaining considerable attentions and establishment of biogas facilities is fast becoming widespread in the continent while issues of energy security and economic growth are also being discussed in several scientific gatherings [32].

3.2 Biobutanol

This is a second generation biofuel produced as a credible substitute for fossil fuel and usually used as a blend with gasoline. Although butanol is still generated through petrochemical methods, the high demand, depletion rate and price of oil has driven the search for a sustainable source for butanol production. This fuel possess some better attributes which includes higher energy content, lower Reid vapor pressure, easy blending with gasoline at any ratio and ease in transportation when compared to bioethanol [27].

3.3 Bioethanol

This is a first generation biofuel mainly produced via enzymatic fermentation by using yeast to digest biodegradable raw materials with high energy content. Hydrolysis is employed when raw materials such as high energy yielding crops are utilized; this is done to break down the complex nature of the polymer into monomers such as simple sugar followed by conversion of the sugar to alcohol after which distillation and dehydration are used to reach the desired amount that can be utilized directly as fuel [33]. Ethanol can be mixed with petrol if appropriately purified and when utilized in modified spark ignition engines, production of toxic environmental gases will be reduced. A liter of ethanol can yield about three fifths of the energy provided by a liter of gasoline [34].

3.4 Biodiesel

Biodiesel is another example of a first generation biofuel and can be produced directly from vegetable oils and other oleo chemicals via trans-esterification methods or cracking. The possibility of biodiesel replacing fossil fuels as main source for power is one reason for the global research of biodiesel [35]. The trans-esterification procedure may utilize acid, enzymes and alcohol to yield the biodiesel and glycerin as by-product [36]. Oleo chemicals are chemical substances produced from fats and natural oils, they are basically fatty acids and glycerol. Hypothetically, oleo chemicals are better substitute for petrochemicals in terms of sustainability and economic viability [37]. The high price rate of biodiesel is a major constraint to its commercialization in contrast with petroleum, thus the utilization of waste oil should be considered since it is relatively available and cheap [38].

4. Biogas development in Sub-Saharan Africa

Biogas generation via anaerobic digestion is very famous in the Americas, Asia, Europe and India Sub-Continent. However, the Sub-Saharan Africa region has over the last few decades witnessed a very slow acceptance and adoption of this technology despite significant individual, institutional, national and international efforts [21]. This slow pace of development has been linked to scarcity or unavailability of feedstock caused by poor agricultural practices [39]. **Table 1** shows that as at 2005, only a few African countries have adopted the biogas technology with an insignificant number of biogas digesters/plants compared to what is obtainable in other continents [15]. In order to improve this situation, a new African initiative was launched in 2007 in order to install biogas digesters to not less than 2 million households by the year 2020 [30, 31]. By the year 2010, the number of biogas plants in Africa has increased especially in Tanzania with about 4000 digester units [40]. However, only about 60% of these plants were functional while the remaining failed or performed below satisfaction due to reasons like planning and construction errors, poor community awareness, lack of adequate maintenance culture, misconception of the technology's benefits, and lack of technical know-how by end-users among others [40].

Country	Number of small/medium digesters (100 m^3)	Number of large digesters (>100 m^3)	Region
Botswana	>100	1	South
Burkina Faso	>30	—	West
Burundi	>279	—	East
Egypt	>100	<100	North
Ethiopia	>100	>1	East
Ghana	>100	—	West
Cote D'Ivoire	>100	1	West
Kenya	>500	—	East
Lesotho	40	—	South
Malawi	—	1	South
Morocco	>100	—	North
Nigeria	Few	—	West
Rwanda	>100	>100	East
Senegal	>100	—	West
Sudan	>200	—	North
South Africa	>100	>100	South
Swaziland	>100	—	South
Tanzania	>1000	1	East
Tunisia	>40	—	North
Uganda	Few	—	East
Zambia	Few	—	East
Zimbabwe	>100	1	South

Source: Mshandete and Parawira [15].

Table 1.
African countries with biogas producing digesters.

5. The Nigeria scenario

Inadequate energy supply and environmental pollution are some of the challenges being faced in Nigeria and other developing nations. The energy consumption rate of the modern world is an indication that renewable and environmental-friendly energy need be generated from alternative sources. The mono digestion of substrates has been found to be limited in both quantity and quality of generated gas while co-digestion of substrates enhance the anaerobic digestion process as this leads to higher carbon/nitrogen balance and nutrient availability. Biogas research in Nigeria is in its infancy as limited substrates have been utilized and significant effort has not been directed at evaluating the composition and/or succession of the microbes responsible for the bioconversions [41]. As seen in **Table 2**, most of the previous biogas researches utilized animal dung, poultry droppings, peels, human

S/N	Substrate	Average biogas/ methane yield	Digestion type	Digestion scale	Reference
1.	Food waste and human excreta	56.5 L/kg biogas	Anaerobic	Pilot	[38]
2.	Poultry dropping	54 L/kg (biogas): 33.3 L/kg (methane)	Anaerobic	Pilot	[73]
3.	*Cymbopogon citratus* and poultry dropping	39 L/kg (biogas): 25.8 L/kg (methane)	Anaerobic	Pilot	[73]
4.	*Cymbopogon citratus*	28 L/kg (biogas): 21.6 L/kg (methane)	Anaerobic	Pilot	[73]
5.	Rice husks	25.1 L/kg (biogas): 21.3 L/kg (methane)	Anaerobic	Pilot	[74]
6.	Cow dung	61.8 L/kg (biogas): 54.2 L/kg (methane)	Anaerobic	Pilot	[75]
7.	*Tithonia diversifolia*	51.8 L/kg (biogas): 40.2 L/kg (methane)	Anaerobic	Pilot	[67]
8.	*Chromolaena odorata* and poultry dropping	64.8 L/kg (biogas): 56.7 L/kg (methane)	Anaerobic	Pilot	[69]
9.	*Tithonia diversifolia* and poultry dropping	61.8 L/kg (biogas): 54.2 L/kg (methane)	Anaerobic	Pilot	[72]
10.	*Arachis hypogeae*	46.8 L/kg (biogas): 38.9 L/kg (methane)	Anaerobic	Pilot	[70]
11.	*Arachis hypogeae* and poultry manure	59.3 L/kg (biogas): 46.6 L/kg (methane)	Anaerobic	Pilot	[68]
12.	*Carica papaya*	58.4 L/kg (biogas): 45.8 L/kg (methane)	Anaerobic	Pilot	[71]
13.	*Carica papaya* and poultry manure	60.1 L/kg (biogas): 54.3 L/kg (methane)	Anaerobic	Pilot	[65]
14.	*Telfairia occidentalis*	46.4 L/kg (biogas): 32.2 L/kg (methane)	Anaerobic	Pilot	[66]
15.	Banana and plantain peels	49.7 L/kg (biogas): 36.2 L/kg (methane)	Anaerobic	Pilot	[51]
16.	*Panicum maximum* and animal wastes	53.4 L/kg (biogas): 42.4 L/kg (methane)	Anaerobic	Pilot	[76]

Table 2.
Previous substrates used for biogas generation in Nigeria.

excreta, agricultural residues and kitchen wastes as feedstock substrates [41–49]. The use of succulent plants for biogas production has been limited to water lettuce, water hyacinth, cassava leaves, *Cymbopogon citratus* and *Eupatorium odoratum* [41–44, 50, 51]. Besides, the detail analysis of lignocellulosic component and optimization of biogas production processes and parameters are lacking in the Nigerian energy literature.

5.1 Biogas technology adoption in Nigeria

Biogas technology's adoption and operation in Nigeria is still at the infancy stage. This slow pace which is similar to the situation in some other Sub-Saharan African countries is caused by unfavorable government policies, inadequate funding of technology and individual's unwillingness [52]. To this end, several feedstocks which are economically suitable for biogas generation in Nigeria have been selectively identified. These include aquatic plants like water lettuce and water hyacinth; agricultural wastes like cow and piggery dung, poultry droppings and processing waste; industrial wastes like municipal solid wastes and sewage [41–43]. Also, the continuous assessment of other locally available materials for their use in biogas production has been made [44]. The use of succulent plants has been limited to water lettuce, water hyacinth, cassava leaves, *Eupatorium odoratum* and *Cymbopogon citratus* [45, 53]. Similarly, the potential of poultry droppings, cow dung and kitchen/food wastes for biogas generation has been experimented upon [54, 55].

6. Suitable feedstock for biogas generation in Sub-Saharan Africa

One of the major steps in achieving anaerobic digestion success is the careful selection and identification of viable feedstock. The world over, several feedstock have been utilized including food wastes, animal dungs, agricultural and plant residues, wastewaters, Organic Fraction of Municipal Solid Wastes (OFMSW), energy crops, etc. Across Sub-Saharan Africa, substrates suitable for anaerobic digestion include aquatic plants such as water lettuce and water hyacinth; agricultural wastes/residues such as cow and piggery dung, *Cymbopogon citratus*, cassava leaves; municipal wastes such as human excreta, processing wastes, urban refuse and industrial wastes [42–46]. Among these, the potentials of poultry manure, cow dung and kitchen wastes for biogas production have been demonstrated [54–59].

Similarly, Ilori et al. [51] demonstrated the biogas generation from the co-digestion of the peels of banana and plantain and obtained the highest gas volume with an equal mass of both substrates. In another study, the co-digestion of pig waste and cassava peels seeded with wood ash produced a significant increase in biogas yield when compared with the unseeded mixture of the substrates [60]. Fariku and Kidah [61] have also reported the efficient generation of biogas from the anaerobic digestion of *Lophira lanceolata* fruit shells. The biogas producing potentials of Sub-Saharan African local algal biomass has been recognized by Weerasinghe and Naqvi [62]. Odeyemi [50] in his comparative study of four substrates (*Eupatorium odoratum*, water lettuce, water hyacinth and cow dung) as potential substrates for biogas production concluded that *Eupatorium odoratum* was the best while cow dung was the poorest substrate in terms of gas yield. Ahmadu [63] compared the biogas production from cow dung and chicken droppings while Igboro [64] compared the biogas from cow dung from an abattoir and the National Animal Production Institute, Zaria, with the abattoir waste generating the highest volume of gas. Igboro [64] also designed a biogas stove burner which was effectively tested with the biogas produced from cow dung and other feed materials.

Recently, there has been an upsurge in the utilization of many novel materials for biogas generation across Sub-Saharan Africa especially in Nigeria and other countries. These biomasses are found abundantly across the region with very little documentations for use as biofuel feedstock. They include shoots of *Tithonia diversifolia* (Mexican sunflower), and *Chromolaena odorata* (Siam weed). Others are fruit peels of *Carica papaya* (pawpaw), *Telfairia occidentalis* (fluted pumpkin), *Ananas comosus* (pineapple), *Citrullus lanatus* (water melon), *Cucumeropsis mannii* (melon) and the hull or pod of *Arachis hypogaea* (peanut or groundnut), *Theobroma cacao* (Cocoa) and *Kola nitida* (kolanut) [14, 65–72]. Despite the huge availability of these biomasses in their various locations of production, they mostly end up as solid wastes in the environment as little or no usage has been sought for them over the years. Even when some of the biomass has been experimented on for biofuel production, the various arrays of microorganisms involved in their biodegradation are yet to be documented in biofuel literature.

7. Conclusion

Sub-Saharan African region is much blessed with diverse biomass and materials that can be exploited for biofuels generation. It has been seen that biofuels especially biogas technology adoption in the region has been slow thereby requiring more concerted efforts. With the past and anticipated energy challenges attributed to the region due to the overdependence on fossil fuels, the generation of environmental friendly biofuels from the locally available biomass in the region should be given top priority as this will help salvage the menace of energy unavailability and its attendant issues.

Acknowledgements

The authors appreciate the support of the technical staff.

Conflicts of interest

Authors declare no conflict of interest.

Funding

This work received funding from Ton Duc Thang University, Ho Chi Minh City, Vietnam.

Author details

Olatunde Samuel Dahunsi[1,2]*, Ayoola Shoyombo[3] and Omololu Fagbiele[4]

1 Faculty of Environment and Labour Safety, Ton Duc Thang University, Ho Chi Minh City, Vietnam

2 Biomass and Bioenergy Group, Environment and Technology Research Cluster, Landmark University, Nigeria

3 Department of Animal Science, Landmark University, Omu-Aran, Kwara State, Nigeria

4 Department of Chemical Engineering, Covenant University, Ota, Ogun State, Nigeria

*Address all correspondence to: dahunsi.olatunde.samuel@tdt.edu.vn

References

[1] Wei S, Zhang H, Cai X, Jin X, Fang J, Liu H. Psychrophilic anaerobic co-digestion of highland barley straw with two animal manures at high altitude for enhancing biogas production. Energy Conversion and Management. 2014;**88**:40-48

[2] Jain S, Jain S, Wolf IT, Lee J, Tong JW. A comprehensive review on operating parameters and different pretreatment methodologies for anaerobic digestion of municipal solid waste. Renewable and Sustainable Energy Reviews. 2015;**52**:142-154

[3] Chirambo D. Addressing the renewable energy financing gap in Africa to promote universal energy access: Integrated renewable energy financing in Malawi. Renewable and Sustainable Energy Reviews. 2016;**62**:793-803

[4] Ge X, Xu F, Li Y. Solid state anaerobic digestion of lignocellulosic biomass: Recent progress and perspectives. Bioresource Technology. 2016;**205**:239-249

[5] Kamp LM, Forn EB. Ethiopia's emerging domestic biogas sector: Current status, bottlenecks and drivers. Renewable and Sustainable Energy Reviews. 2016;**60**:475-488

[6] Mengistu MG, Simane B, Eshete G. Factors affecting households' decisions in biogas technology adoption, the case of Ofla and Mecha Districts, northern Ethiopia. Renewable Energy. 2016;**93**:215-227

[7] Mungwe JN, Colombo E, Adani F, Schievano A. The fixed dome digester: An appropriate design for the context of Sub-Saharan Africa? Biomass and Bioenergy. 2016;**95**:35-44

[8] Wang Y, Zhu G, Song L, Wang S, Yin C. Manure fertilization alters the population of ammonia-oxidizing bacteria rather than ammonia-oxidizing archaea in a paddy soil. Journal of Basic Microbiology. 2013;**100**:1-8

[9] Zou S, Wang X, Chen Y, Wan H, Feng Y. Enhancement of biogas production in anaerobic co-digestion by ultrasonic pretreatment. Energy Conversion and Management. 2016;**112**:226-235

[10] Abadi N, Gebrehiwot K, Techane A, Nerea H. Links between biogas technology adoption and health status of households in rural Tigray, Northern Ethiopia. Energy Policy. 2017;**101**:284-292

[11] Ohimain EI, Izah SC. A review of biogas production from palm oil mill effluents using different configurations of bioreactors. Renewable and Sustainable Energy Reviews. 2017;**70**:242-253

[12] Roopnarain A, Adeleke R. Current status, hurdles and future prospects of biogas digestion technology in Africa. Renewable and Sustainable Energy Reviews. 2017;**67**:1162-1179

[13] Russo V, von Blottnitz H. Potentialities of biogas installation in South African meat value chain for environmental impacts reduction. Journal of Cleaner Production. 2017;**153**:465-473

[14] Shane A, Gheewala SH, Kafwembe Y. Urban commercial biogas power plant model for Zambian towns. Renewable Energy. 2017;**103**:1-14

[15] Mshandete AM, Parawira W. Biogas technology research in selected Sub Saharan Africa. African Journal of Biotechnology. 2009;**8**(2):116-125

[16] Valentine J, Clifton-Brown J, Hastings A, Robson P, Allison G, Smith P. Food vs. fuel: The use of land for

lignocellulosic 'next generation' energy crops that minimize competition with primary food production. GCB Bioenergy. 2012;**4**(1):1-19

[17] Khoufi S, Louhichi A, Sayadi S. Optimization of anaerobic co-digestion of olive mill wastewater and liquid poultry manure in batch condition and semi continuous jet-loop reactor. Bioresource Technology. 2015;**182**:67-74

[18] Giwa A, Alabi A, Yusuf A, Olukan T. A comprehensive review on biomass and solar energy for sustainable energy generation in Nigeria. Renewable and Sustainable Energy Reviews. 2017;**69**:620-641

[19] Calabro PS, Greco R, Evangelou A, Komilis D. Anaerobic digestion of tomato processing waste: Effect of alkaline pretreatment. Journal of Environmental Management. 2015;**163**: 49-52

[20] Yasar A, Rasheed R, Tabinda AB, Tahir A, Sarwar F. Life cycle assessment of a medium commercial scale biogas plant and nutritional assessment of effluent slurry. Renewable and Sustainable Energy Reviews. 2017;**67**:364-371

[21] Lynd LR, Sow M, Chimphango AFA, Cortez LAB, Cruz CHB, Elmissiry M, et al. Bioenergy and African transformation. Biotechnology for Biofuels. 2015;**8**(18):1-18

[22] Su H, Liu L, Wang S, Wang Q, Jiang Y, Hou X, Tan T. Semi continuous anaerobic digestion for biogas production: Influence of ammonium acetate supplement and structure of the microbial community. Biotechnology for Biofuels. 2015;**8**(13):1-13

[23] Kwietniewska E, Tys J. Process characteristics, inhibition factors and methane yields of anaerobic digestion process, with particular focus on microalgal biomass fermentation. Renewable and Sustainable Energy Reviews. 2014;**34**:491-500

[24] Sawasdee V, Pisutpaisal N. Feasibility of biogas production from Napier grass. Energy Procedia. 2014;**61**:1229

[25] Petersson A, Thomsen MH, Hauggaard-Nielsen H, Thomsen AB. Potential bioethanol and biogas production using lignocellulosic biomass from winter rye, oilseed rape and faba bean. Biomass and Bioenergy. 2007;**31**:812-819

[26] Morone A, Pandey RA. Lignocellulosic biobutanol production: Gridlocks and potential remedies. Renewable and Sustainable Energy Reviews. 2014;**37**:21-35

[27] Larson ED. Biofuel production technologies: Status, prospects and implications for trade and development. Report No. UNCTAD/DITC/TED/2007/10. New York and Geneva: United Nations Conference on Trade and Development; 2008

[28] Ezeonu CS, Ezeonu NC. Alternative sources of petrochemicals from readily available biomass and agro-products in Africa: A review. Journal of Petroleum and Environmental Biotechnology. 2016;**7**(5):12-23

[29] Amigun B, Sigamoney R, Von Blottnitz H. Commercialization of biofuel industry in Africa: A review. Renewable and Sustainable Energy Reviews. 2008;**12**:690-711

[30] Adeniyi OD, Kovo AS, Abdulkareem AS, Chukwudozie C. Ethanol fuel production from cassava as a substitute for gasoline. Dispersion and Technology Journal. 2007;**28**:501-504

[31] Ayhan D. Importance of biomass energy sources for Turkey. Energy Policy Journal. 2008;**36**:834-842

[32] Soumonni O, Cozzens S. The potential for biofuel production and use in Africa: An adaptive management approach. In: VI Globelics Conference; Mexico City. 2008

[33] IEA. Biofuels for transporte an international perspective. Paris, France: International Energy Agency (IEA); 2004. http://www.iea.org/textbase/nppdf/free/2004/biofuels2004.pdf

[34] Barakat A, Monlau F, Solhy A, Carrere H. Mechanical dissociation and fragmentation of lignocellulosic biomass: Effect of initial moisture, biochemical and structural proprieties on energy requirement. Applied Energy. 2015;**142**:240-246

[35] Owolabi RU, Adejumo AL, Aderibigbe AF. Biodiesel: Fuel for the future (a brief review). International Journal of Energy Engineering. 2012;**2**:223-231

[36] Nigram PS, Singh A. Production of liquid biofuels from renewable resources. Progress in Energy and Combustion Science. 2011;**37**:52-68

[37] Naik SN, Goud VV, Rout PK, Dalai AK. Production of first and second generation biofuels: A comprehensive review. Renewable and Sustainable Energy Reviews. 2010;**14**:578-597

[38] Zhang Y, Dube MA, McLean DD, Kates M. Biodiesel production from waste cooking oil: Economic assessment and sensitivity analysis. Bioresource Technology. 2003;**90**:229-240

[39] USDA/FAS. World Report: Cattle Population by Country. United States Department of Agriculture/Foreign Agricultural Service. United States; 2008

[40] Ocwieja SM. Life Cycle Thinking Assessment Applied to Three Biogas Projects in Central Uganda, Being a Report Submitted in Partial Fulfillment of the Requirements for the Degree of Master of Science in Environmental Engineering. United States: Michigan Technological University; 2010

[41] Akinbami JFK, Akinwumi IO, Salami AT. Implications of environmental degradation in Nigeria. Natural Resource Forum. 1996;**20**:319-331

[42] Akinbami JFK, Ilori MO, Oyebisi TO, Akinwuni IO, Adeoti O. Biogas energy use in Nigeria: Current status, future prospects and policy implications. Renewable, Sustainable Energy Review. 2001;**5**:97-112

[43] Okagbue RN. Fermentation research in Nigeria. MIRCEN Journal. 1988;**4**:169-182

[44] Ubalua AO. Cassava wastes: Treatment options and value addition alternatives. African Journal of Biotechnology. 2008;**6**:2065-2073

[45] Alfa IM, Okuofu CA, Adie DB, Dahunsi SO, Oranusi US, Idowu SA. Evaluation of biogas potentials of *Cymbopogon Citratus* as alternative energy in Nigeria. International Journal of Green Chemistry and Bioprocess. 2012;**2**(4):34-38

[46] Dahunsi SO, Oranusi US. Co-digestion of food waste and human excreta for biogas production. British Biotechnology Journal. 2013;**3**(4):485-499

[47] Adepoju TF, Eyibio UP, Olatunbosun B. Optimization investigation of biogas potential of *Tithonia diversifolia* as an alternative energy source. International Journal of Chemical and Process Engineering Research. 2016;**3**(3):46-55

[48] Ibrahim MD, Imrana G. Biogas production from lignocellulosics materials: Co-digestion of corn cobs, groundnut shell and sheep dung. Imperial Journal of Interdisciplinary Research. 2016;**2**(6):5-11

[49] Idire SO, Asikong BE, Tiku DR. Potentials of banana peel, vegetable waste (*telfairia occidentalis*) and pig dung substrates for biogas production. British Journal of Applied Science and Technology. 2016;**16**(5):1-6

[50] Odeyemi O. Biogas from *Eupatorium odorantum*, an alternative cheap energy source for Nigeria. In: Emejuaiwe SO, Ogunbi O, Sanni SO, editors. Global impacts of Applied Microbiology, 6th International Conference. London: Academic Press; 1981. pp. 246-252

[51] Ilori OM, Adebusoye AS, Lawal AK, Awotiwon AO. Production of biogas from banana and plantain peels. Advances in Environmental Biology. 2007;**1**(1):33-38

[52] Sokoto Energy Research Centre. Information brochure on biogas generation and utilization. Usmanu Danfodiyo University, Sokoto; 2004

[53] Odeyemi O. Resource assessment for biogas production in Nigeria. Nigerian Journal of Microbiology. 1983;**3**:59-64

[54] Lawal AK, Ayanleye TA, Kuboye AO. Biogas production from some animal wastes. Nigerian Journal of Microbiology. 1995;**10**:124-130

[55] Ojolo SJ, Dinrifo RR, Adesuyi KB. Comparative study of biogas production from five substrates. Advanced in Materials Research Journal. 2007;**18**(19):519-525

[56] Matthew P. Gas production from animal wastes and its prospects in Nigeria. Nigerian Journal of Solar Energy. 1982;**2**(98):103-109

[57] Akinluyi TO, Odeyemi O. Comparable seasonal methane production of five animal manures in Ile-Ife, Nigeria. In: Abstracts, 14th Annual Conference, Nigerian Society for Microbiology. 1986. p. 5

[58] Abubakar MM. Biogas generation from animal wastes. Nigerian Journal of Renewable Energy. 1990;**1**:69-73

[59] Zuru AA, Saidu H, Odum EA, Onuorah OA. A comparative study of biogas production from horse, goat and sheep dung. Nigerian Journal of Renewable Energy. 1998;**6**:43-47

[60] Adeyanju AA. Effect of seeding of wood-ash on biogas production using pig waste and cassava peels. Journal of Engineering and Applied Sciences. 2008;**3**:242-245

[61] Fariku S, Kidah MI. Biomass potentials of *Lophira lanceolata* fruit as a renewable energy resource. African Journal of Biotechnology. 2008;**7**:308-310

[62] Weerasinghe B, Naqvi SHZ. Algal bioconversion of solar energy to biogas for rural development in the Sub-Saharan region. In: Paper presented at the Science Association of Nigeria Conference; Ibadan. 1983

[63] Ahmadu TO. Comparative performance of cow dung and chicken droppings for biogas production [M.Sc thesis]. Zaria: Department of Mechanical Engineering, Ahmadu Bello University; 2009

[64] Igboro SB. Production of Biogas and Compost from Cow Dung in Zaria, Nigeria. In: Presented to the Department of Water Resources and Environmental Engineering [unpublished PhD dissertation]. Zaria, Nigeria: Ahmadu Bello University; 2011

[65] Dahunsi SO, Oranusi S, Owolabi JB, Efeovbokhan VE. Mesophilic anaerobic co-digestion of poultry droppings and *Carica papaya* peels: Modelling and process parameter optimization study. Bioresource Technology. 2016;**216**:587-600

[66] Dahunsi SO, Oranusi S, Owolabi JB, Efeovbokhan VE. Comparative biogas

generation from fruit peels of fluted pumpkin (*Telfairia occidentalis*) and its optimization. Bioresource Technology. 2016;**221**:517-525

[67] Dahunsi SO, Oranusi S, Efeovbokhan VE. Anaerobic mono-digestion of *Tithonia diversifolia* (wild Mexican sunflower). Energy Conversion and Management. 2017;**148**:128-145

[68] Dahunsi SO, Oranusi S, Efeovbokhan VE. Pretreatment optimization, process control, mass and energy balances and economics of anaerobic co-digestion of *Arachis hypogaea* (peanut) hull and poultry manure. Bioresource Technology. 2017;**241**:454-464

[69] Dahunsi SO, Oranusi S, Owolabi JB, Efeovbokhan VE. Synergy of Siam weed (*Chromolaena odorata*) and poultry manure for energy generation: Effects of pretreatment methods, modeling and process optimization. Bioresource Technology. 2017;**225**:409-417

[70] Dahunsi SO, Oranusi S, Efeovbokhan VE. Optimization of pretreatment, process performance, mass and energy balance in the anaerobic digestion of *Arachis hypogaea* (peanut) hull. Energy Conversion and Management. 2017;**139**:260-275

[71] Dahunsi SO, Oranusi S, Efeovbokhan VE. Cleaner energy for cleaner production: Modeling and optimization of biogas generation from *Carica papayas* (pawpaw) fruit peels. Journal of Cleaner Production. 2017;**156**:19-29

[72] Dahunsi SO, Oranusi S, Efeovbokhan VE. Bioconversion of *Tithonia diversifolia* (Mexican sunflower) and poultry droppings for energy generation: Optimization, mass and energy balances, and economic benefits. Energy and Fuels. 2017;**31**:5145-5157

[73] Owamah HI, Alfa MI, Dahunsi SO. Optimization of biogas from chicken droppings with *Cymbopogon citratus*. Renewable Energy. 2014;**68**:366-371

[74] Okeh OC, Onwosi OC, Odibo FJ. Biogas production from rice husks generated from various rice mills in Ebonyi State Nigeria. Renewable Energy. 2013;**62**:204-208

[75] Ahmadu TO, Folayan CO, Yawas DS. Comparative performance of cow dung and chicken droppings for biogas production. Nigerian Journal of Engineering. 2009;**16**(1):154-164

[76] Uzodinma EO, Ofoefule AU. Biogas production from blends of field grass (*Panicum maximum*) with some animal wastes. International Journal of Physical Sciences. 2009;**4**(2):091-095